EARTHLIGHT

REFLECTIONS FROM SPACE:
THE POWER OF EARTHLIGHT AND THE HUMAN PERSPECTIVE

Dr. Sian Proctor

CONTENTS

Foreword

On September 15, 2021, the historic Inspiration4 private spaceflight launched into outer space. This was the first all-civilian orbital spaceflight, a mission that ushered in a new era of human space travel. On that mission, Dr. Sian Proctor was one of the four astronauts aboard the SpaceX Dragon. She was given the call sign, "Leo."

There are many reasons this call sign would have been a good choice. In astrology and astronomy, for example, Leo is the constellation of the lion, a sign of leadership. Sian is, indeed, a leader. She has been first in many areas, including becoming the first African American woman to be the pilot on a space mission. "Leo" could also refer to Low Earth Orbit (LEO), which was where the crew would be going.

When I found out the real origins of the call sign, it seemed even more appropriate: Sian was nicknamed "Leo" for Leonardo da Vinci, the original Renaissance person. He was a scientist, engineer, and artist, a man of multiple interests and talents. Sian shares many personality characteristics with da Vinci. He was obsessed with human flight, and worked hard at making it happen, even though he was never able to achieve it. Da Vinci was gracious, he was elegant, and expressed himself with color. But most notably, they share an unquenchable desire for knowledge.

Sian's education is in the geosciences; she is a professor at Maricopa Community Colleges in that field. However, when she competed for a seat on Inspiration4, she did so as an artist. Sian fulfilled the dream of her astronaut predecessors that we would send poets and artists into space. They knew an artist, like Sian, could share the experience better than any test pilot or engineer.

While on orbit, Sian created works of art to commemorate her experience, following in the footsteps of her good friend, Nicole Stott, the first astronaut to paint in outer space.

In addition to these firsts, I would like to add another one that seems unique to me personally. I have interviewed some 50 astronauts about their experiences of spaceflight, and I believe that Sian was the first to talk about "Earthlight."

We know that "Moonlight" is a result of the sun's rays being reflected by the surface of the Moon. However, those of us confined to the surface of the planet cannot truly perceive "Earthlight," in the same way that we experience Moonlight. That is an experience exclusive to those who leave Earth's atmosphere and look back at our beautiful home in its place among the stars.

Other astronauts have shared many wonderful stories with me about the Overview Effect perspective, including that the Earth "just glows," that they see incredible colors from orbit, the borders and boundaries between countries simply fade away, and the blue line of the atmosphere, that keeps us alive, is incredibly thin. Only Sian, though, has told me that she was "bathed in Earthlight" when she floated into the Dragon Capsule's giant cupola window.

So let's turn to a discussion of the book itself. Most books are one thing or another: they are scientific books, poetry books, art books, or something else altogether. Because "Leo" is a Renaissance person, though, Earthlight is all of these.

Laid out beautifully by the talented husband-and-wife team of John and Jennifer Read, the book includes a scientific explanation of Sunlight, Moonlight, and Earthlight. Perhaps most important is this: the author not only includes the science, but it is also understandable to non-scientists!

In addition, Sian includes her artwork, some of which was produced in orbit. She also includes her poetry, which is surely destined to produce a new genre: Earthlight poems.

As the author of The Overview Effect: Space Exploration and Human Evolution, I was delighted to see an entire chapter on the Overview Effect, which is closely related to Earthlight. This chapter includes numerous quotes from astronauts sharing their experience with the phenomenon. The book even links Earthlight to highly practical terrestrial concerns, like climate change.

Overall, Earthlight is a remarkable achievement, interweaving "Leo's" talents and interests into an informative, inspiring, and delightful read. It represents a major contribution to the literature of spaceflight.

Frank White, Author of The Overview Effect: Space Exploration and Human Evolution, January 2024

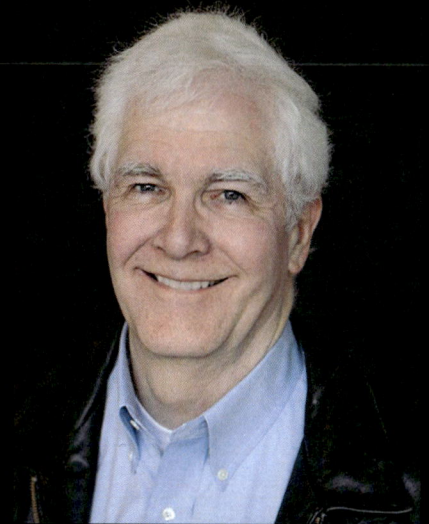

Top 10 Benefits of Learning about Earthlight

1. Understanding Physics:

Earthlight offers a unique opportunity to learn about the fundamental physics of light, including reflection, absorption, and scattering.

2. Appreciation of Earth:

Understanding Earthlight increases our appreciation of our planet. Without sunlight transformed into Earthlight, life wouldn't exist.

3. Astronaut's Perspective:

Exploring Earthlight helps us understand the astronaut's experience, and the psychological and emotional impact known as the 'overview effect'.

4. Climate Studies:

Earthlight studies help us to understand climate science, as changes in Earth's albedo are closely linked to climate change.

5. Astronomy:

Earthlight is a great entry point to learn about astronomy and space exploration.

6. Understanding Moonlight:

Understanding Earthlight is integral to understanding Moonlight and why the lunar phases appear as they do from Earth.

7. Cultural Connection:

Moonlight and Earthlight have been part of human stories, myths, and legends for millennia. Understanding these phenomena can enhance our appreciation of human culture.

8. Interdisciplinary Knowledge:

Earthlight sits at the intersection of physics, astronomy, psychology, and philosophy, helping to bridge these disciplines.

9. Cosmic Perspective:

Earthlight gives us a broader perspective about our place in the universe.

10. Space Art and Photography:

Understanding Earthlight improves space art and photography.

View from ISS

AfroGaia Rising by Dr. Sian Proctor

BETWEEN EARTH AND SPACE IS SKY

The Transformative Power of Earthlight

EarthLight

I thought the moonlight was my guiding light
Until that day when my soul shimmered
Eyes wide and dilated with realization
For there I was being bathed in Earthlight

Tasered by the pulsating Earth glow
My weeping ego quivers
Spellbound in awe at the cosmic chaos
Perched against the death

A clear beacon of hope and longing
Etched by complex molecules and spiraling DNA
Golden strands of energy cascading outward
Encapsulating hopes and dreams
Existence and affirmation

The baby's blanket ripped away
I howl at the sensation
Love struck in suspension

My mind struggles to comprehend
So much meta transcending time and space
Who will hear the cries of the generations
AfroGaia simmers under the weight of memories

I hold court among the stars & testify to the cosmos
All our hopes set adrift
Let us be set free in a sea of forgiveness for what we have not seen

If only we could all be baptized
by Earthlight

Dr. Sian Proctor

NASA Artist Concept of Artemis I

As the Earth spins in the endless expanse of space, it radiates a subtle glow, a testament to the dance of light between our planet, its moon, and the Sun. This phenomenon, known as Earthlight, shapes our understanding of our celestial home and profoundly influences those who have had the chance to witness it firsthand.

Introduction

Welcome to "Reflections from Space: The Power of Earthlight and the Human Perspective," an exploration of the fascinating world of Earthlight and its impact on science, culture, and our worldview.

In the celestial dance of the Sun, the Earth, and the Moon, an extraordinary symphony of light unfolds. This symphony illuminates our understanding of our world, inspires our imagination, and echoes in the silent void of space. Prepare to embark on a journey to better understand our place in the cosmos, appreciate the grandeur of our planet, and witness how the omnipresent yet often overlooked power of light profoundly influences our existence.

Only a few humans have been privileged to see our planet firsthand from space. We will delve into the captivating experience of astronauts who have witnessed the entirety of Earthlight, forever altering their perception of life and our planet, a phenomenon known as the Overview Effect.

"In the celestial dance of the Sun, the Earth, and the Moon, an extraordinary symphony of light unfolds."

and myths.

Earthlight also tells a tale of our changing climate. From aboard dozens of satellites, instruments observe shifts in Earth's reflectivity, offering vital clues to global transformations. Astronaut photographers, like Chris Hatfield, have captured the dynamic changes in Earthlight, portraying once-lush rainforests that now stand barren and the expanding footprint of cities juxtaposed with the dwindling expanses of pristine forests. These images have forever changed our perspective of our home planet.

As you journey through this book, prepare for a new appreciation for the delicate balance and intricate mechanisms governing Earthlight and Moonlight. You'll see our world through the eyes of people profoundly moved by the spectacle of Earth, suspended in a sunbeam. You'll gain a unique understanding of Earthlight and the intricate interplay of light in our corner of the universe.

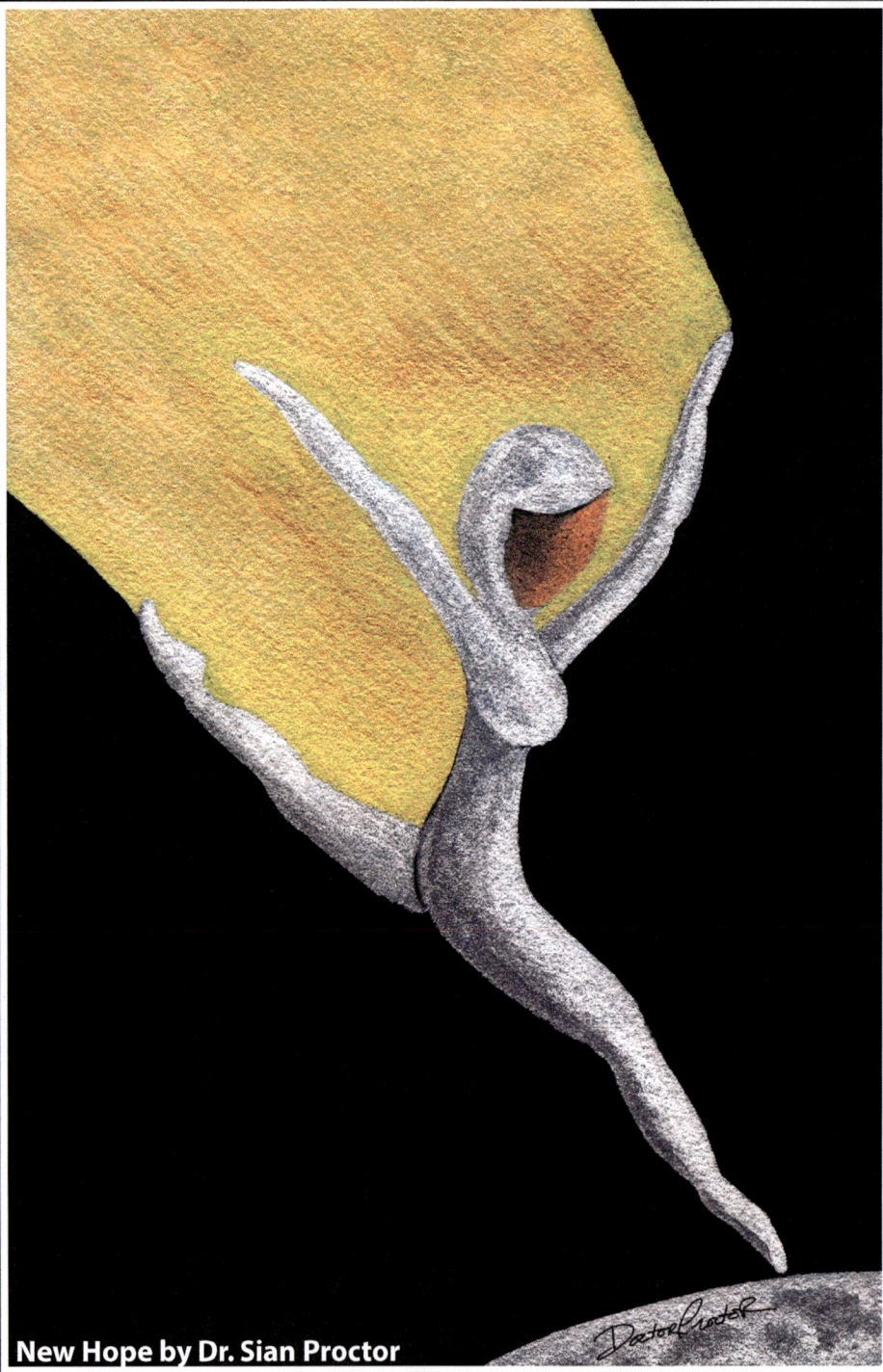

New Hope by Dr. Sian Proctor

Chapter 1

THE DANCE OF LIGHT

Understanding Sunlight, Earthlight, and Moonlight

Dance of Light

The dance of Earth's Light is captivatingly bold,
A ballet of hues for the young and old.

Humanity is mesmerized by
visible light,
Breaking the bonds of our endless night.

Dazzling light beams reflect, scatter, and
intertwine,
A modulating canvas of ever-changing design.

A symphony played for eyes
perched upon the sky,
Whisked through time as
we love, laugh, and cry,

Afronaut by Dr. Sian Proctor

History recorded with beauty and grace,
Pumping the color filled veins of empty space.

A bitterly brief moment marked
by the tick tock of time, as we
majestically swirl within the goldest sublime.

Inhaling the magic of life through the lungs of air,
Revealing a world filled with awe, wonder, and
spectacular flare.

It's a poet that enchants us
with every glance,
The spinning swirl of the
Earth's Light dance.

Have you ever contemplated the grandeur of the universe?

Amongst its boundless wonders, there's a fascinating spectacle happening right in our cosmic backyard: the delicate ballet of light between the Sun, Earth, and Moon. This dance is not merely a silent display of illumination; it molds our perception of our world and the universe around us, giving birth to the mesmerizing phenomena known as Earthlight and Moonlight.

Earthlight isn't just a matter of sunlight hitting the Earth. It's an intricate cascade of processes. Each ray of sunlight that makes the 93 million mile journey across space interacts with our planet's unique features. Beams of Sunlight absorbed, scattered, or reflected, are transformed into Earthlight.

Earth's atmosphere plays a vital role in shaping Earthlight. It acts like a colossal prism, scattering some light waves more than others and infusing the Earthlight with beautiful hues. It's this scattering that paints the sky blue.

Earth's oceans also play a role. They absorb red light and scatter blue, giving our planet its stunning "Blue Marble" appearance when viewed from space.

This dazzling Earthlight doesn't end at the boundaries of our planet; it continues its journey to the Moon. There, the Earthlight undergoes another transformation. Due to the Moon's lack of atmosphere and its different reflective properties, the Earthlight morphs into the serene, gentle Earthshine that we see illuminating the rest of the Moon during its crescent phase just after sunset or before sunrise.

We're not merely observers of this spectacle. We're part of it. The interplay of light between the Sun, Earth, and Moon impacts us in profound ways, from guiding our ancestors to fostering advancements in science.

Our understanding of Earthlight and Moonlight doesn't just deepen our knowledge of the universe—it connects us more intimately with our cosmic home.

Sunlight and the Electromagnetic Spectrum

At the heart of this celestial ensemble is the Sun, our system's radiant star, and life's powerhouse. Emitting an array of electromagnetic radiation and charged particles, the Sun bathes our world in light, sparking the cascade of processes that lead to the creation of Earthlight.

The Sun, a massive fusion reactor that provides the light and heat necessary for life on Earth.

Light is far more than meets the eye

Light plays a fundamental role in how we perceive the world. The visible spectrum that human eyes perceive shares the stage with a range of wavelengths and frequencies we call the electromagnetic spectrum, encompassing everything from X-rays and gamma rays to microwaves and radio waves.

Reflected Solar Radiation
(Visible Light)

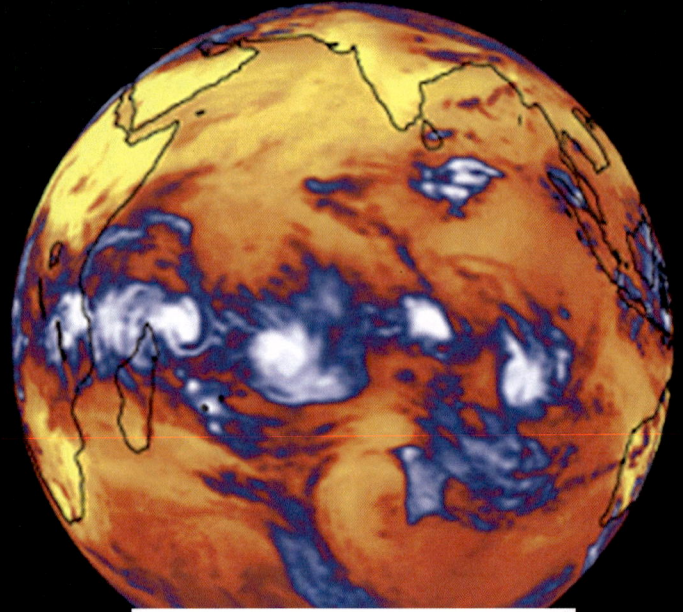

Emitted Heat Radiation
(Infrared Radiation)

Light is the complete electromagnetic spectrum

Imagine a vast cosmic keyboard, with each key representing a different wavelength. At one end, we find the low-frequency (and low energy) notes of radio waves. Moving up the keyboard, we reach the high-frequency (high energy) tones of gamma rays. Our eyes are attuned to play a minuscule portion of this cosmic keyboard – the realm of visible light. "Light" is the complete electromagnetic spectrum.

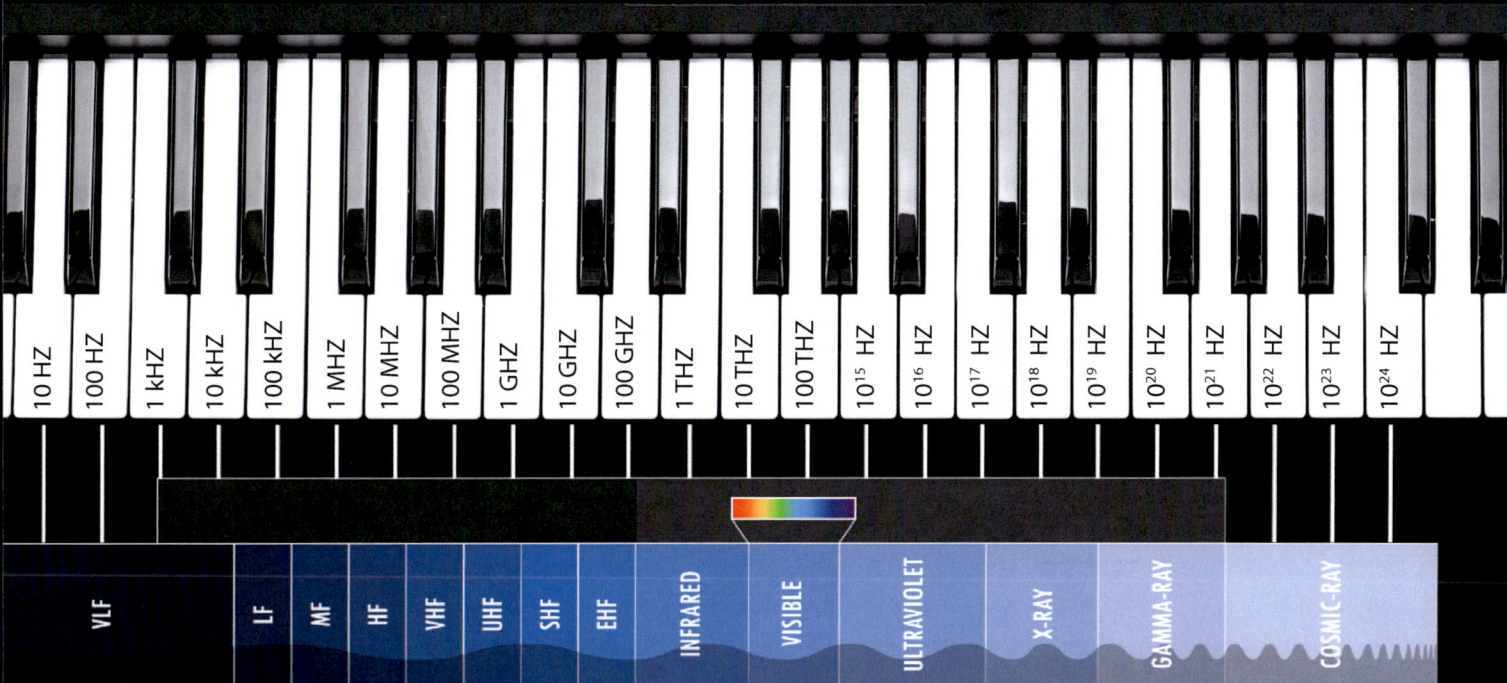

Keyboard key labels: 10 HZ, 100 HZ, 1 kHZ, 10 kHZ, 100 kHZ, 1 MHZ, 10 MHZ, 100 MHZ, 1 GHZ, 10 GHZ, 100 GHZ, 1 THZ, 10 THZ, 100 THZ, 10^{15} HZ, 10^{16} HZ, 10^{17} HZ, 10^{18} HZ, 10^{19} HZ, 10^{20} HZ, 10^{21} HZ, 10^{22} HZ, 10^{23} HZ, 10^{24} HZ

Spectrum bands: VLF, LF, MF, HF, VHF, UHF, SHF, EHF, INFRARED, VISIBLE, ULTRAVIOLET, X-RAY, GAMMA-RAY, COSMIC-RAY

The electromagnetic spectrum is comprised of light waves at different energies, categorized by their frequency and wavelength.

Visible light, an ensemble of different 'notes,' enables us to interpret the world. When it encounters an object, it chooses one of three paths: absorption, reflection, or refraction.

The vibrant colors we perceive are the symphony resulting from these interactions, with each color playing a specific note within the visible light spectrum.

Absorbed light mostly gets turned into heat, but it can also trigger chemical reactions, phosphoresce at different wavelengths (think glow in the dark), or generate electric current like in a solar panel.

TOP OF ATMOSPHERE

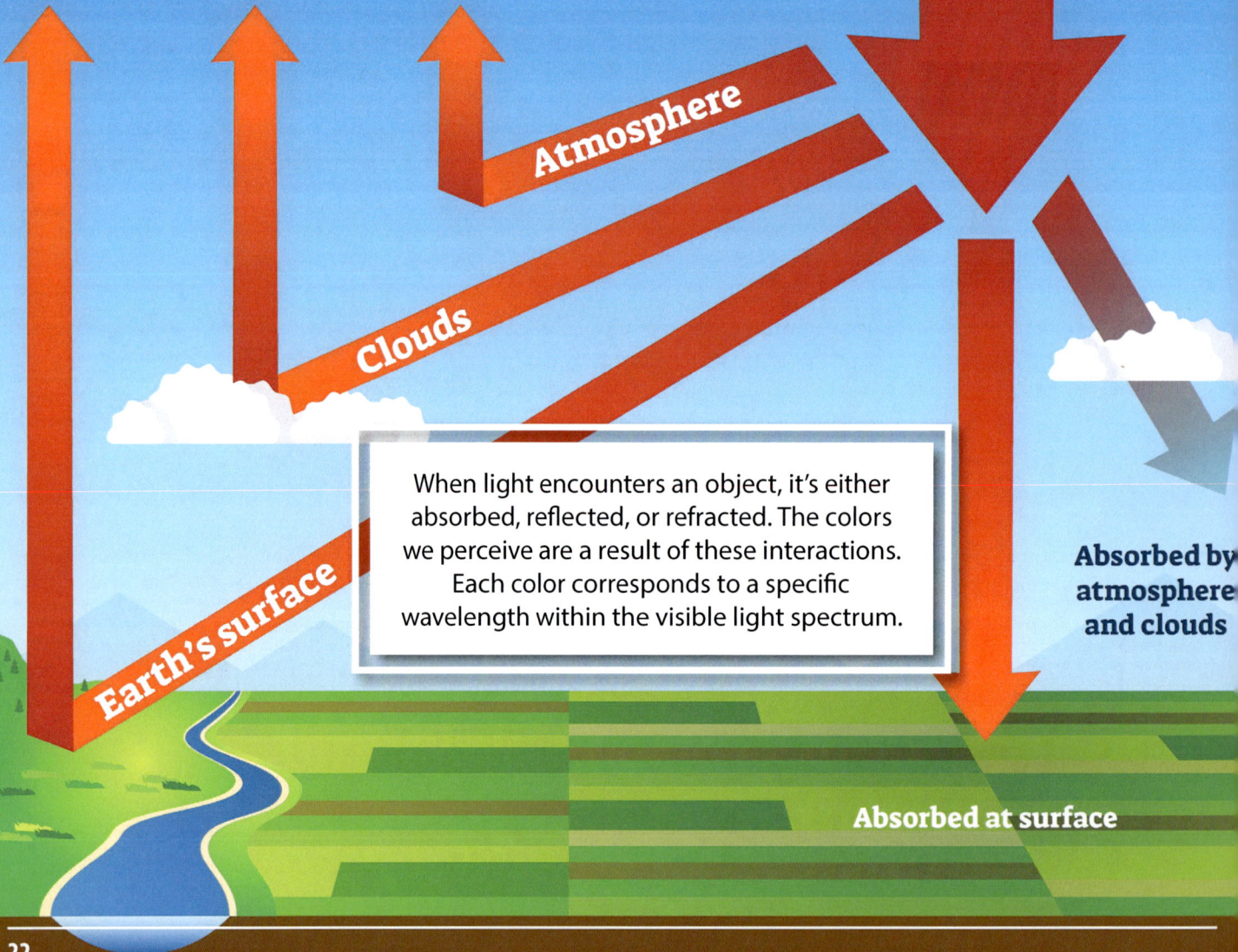

Atmosphere

Clouds

Earth's surface

When light encounters an object, it's either absorbed, reflected, or refracted. The colors we perceive are a result of these interactions. Each color corresponds to a specific wavelength within the visible light spectrum.

Absorbed by atmosphere and clouds

Absorbed at surface

Sunlight, the grand symphony of light, contains all visible colors. When this chorus of colors encounters a prism, it diverges into a rainbow of individual notes, a phenomenon known as dispersion. It's this intricate dance of light, this celestial symphony, that paints the world in colors and underlies the phenomena of Earthlight and Moonlight.

Our planet is the prism. As soon as sunlight strikes our planet, that light immediately changes into Earthlight through the process of absorption, reflection, and refraction.

Sunlight, or white light, is a mix of all colors in the visible spectrum. When it passes through a prism, it splits into a rainbow of colors, an event known as dispersion.

The Sun-Earth-Moon System

In the cosmic theater of our solar system, the Sun, Earth, and Moon enact a captivating celestial ballet. This trio, bound by the invisible threads of gravity and light, stages a dazzling spectacle, giving birth to the phenomena we know as Earthlight and Moonlight.

The distance between the Sun and the Earth-Moon system is 93 million miles. It takes sunlight 8.3 minutes to travel from the Sun to the Earth-Moon system. It takes 1.3 seconds for light to travel between the Earth and Moon.

Earth, our planetary home, plays its part by possessing a distinct set of features. Paramount among these is its atmosphere, a protective veil of gases and plasma. This layer captures, reflects, and scatters sunlight in a complex dance, with a fraction of this reflected radiance setting off on a cosmic journey back to space as Earthlight.

The reflection of sunlight on the Moon is what we see as Moonlight. When the Moon is in its crescent phase, a faint glow illuminating the unlit part of the Moon can be seen from Earth. This glow is Earthshine, a reflection of Earthlight.

Codex Leicester

Leonardo da Vinci was the first to document Earthlight reflecting off the Moon's surface, a phenomenon known as Earthshine.

Earthlight as Observed From Orbit

Reflection:

This is the primary mechanism behind Earthlight that astronauts experience in space. The sunlight hitting the Earth's surface is reflected back into space. The amount of light reflected depends on the albedo, or reflectivity, of the Earth's surface. Snow and ice, for example, have high albedo and reflect more light, whereas water and vegetation have lower albedo and reflect less light.

Scattering:

Sunlight interacts with the Earth's atmosphere, undergoing scattering. This is the same principle that makes the sky look blue (Rayleigh scattering). Scattering can change the direction and the wavelength of light, thereby influencing the color and intensity of Earthlight.

Geometric factors:

The Earth's shape (an oblate spheroid) and its axial tilt affect how sunlight is received and reflected. The position of the observer (on the Moon, in space, etc.) also influences the perception of Earthlight.

Time of year:

The Earth's axial tilt causes seasons, which in turn change the Earth's albedo as ice coverage and vegetation vary through the year.

Atmospheric factors:

The composition of the Earth's atmosphere (cloud cover, pollutants, etc.) affects the properties of Earthlight reflected.

Re-emission:

Some sunlight is absorbed by the Earth's surface and then re-emitted as longer-wavelength light (infrared). However, this is usually not visible to the naked eye, although it's critical in understanding the Earth's energy budget and heat balance.

"Earthlight" refers to the sunlight reflected off the Earth's surface or atmosphere, but it is fundamentally any light that has been transformed by the Earth, whether seen from space or the ground.

Human contributions to Earthlight

Although reflected sunlight is by far the dominant source of Earthlight as seen from space, humans have begun to contribute to the Earth's glow as well, and not always in a good way.

In cities, artificial light reflecting off of particles in the air blocks all but the brightest stars. Even from the suburbs, light trespassing from streetlamps, homes, and businesses has its share of consequences.

Human-induced changes to Earth's atmosphere not only stop infrared light from escaping into space, but also influence cloud cover, which changes how much light reaches Earth's surface.

Humans are also showering the cosmos with electromagnetic radiation all across the spectrum. Though long wavelength signals (like HF radio) bounce off the ionosphere and return to Earth, the higher frequency signals used by Ham Radio operators and television are beamed unimpeded into space, where they continue traveling indefinitely.

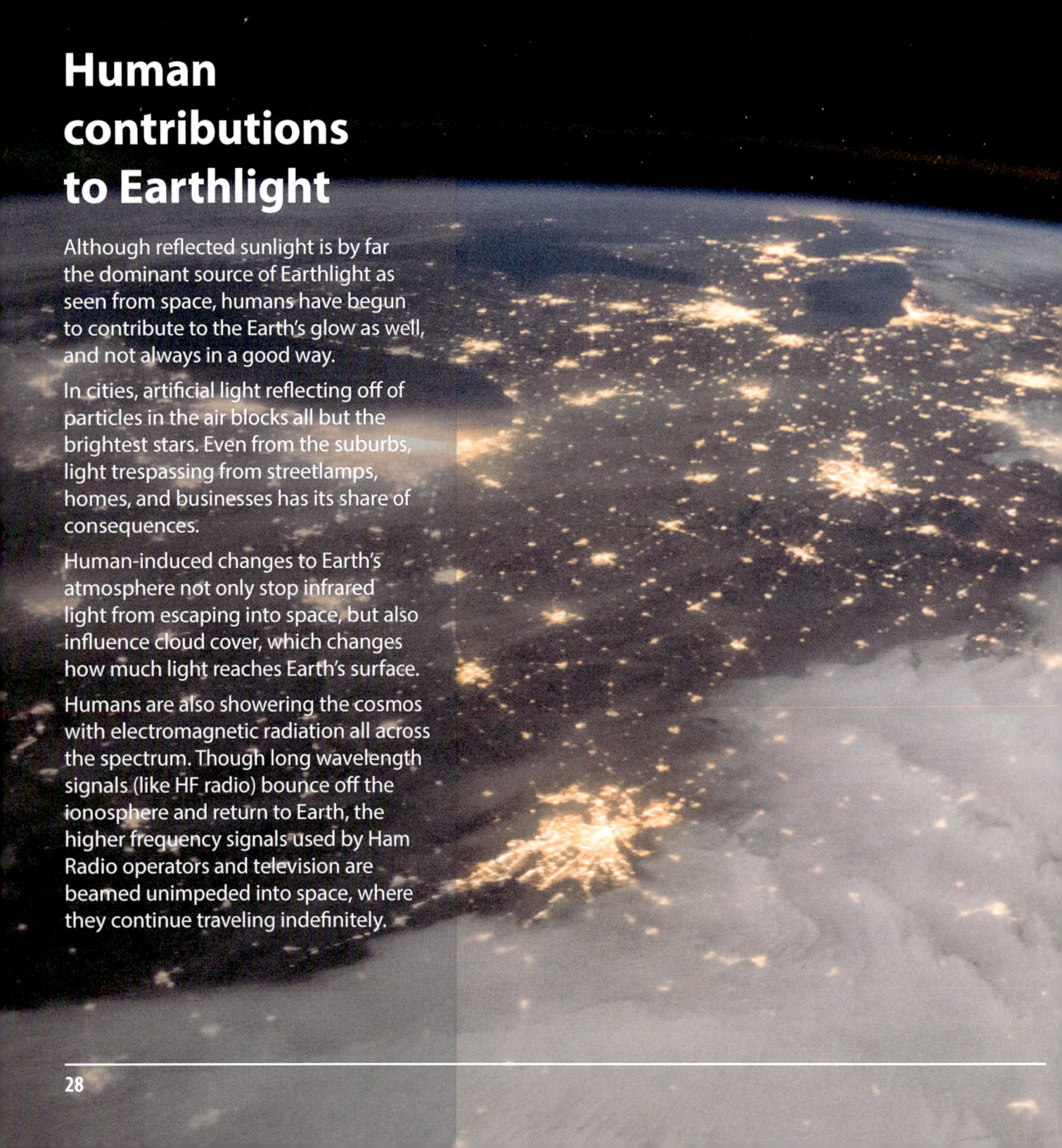

Light Pollution:

Viewing the Earth's nighttime side from space showed our interconnected world awash in lights. A patchwork of cities and towns illuminates nearly every continent.

Light pollution does more than block the stars, it affects our health too. Our bodies rely on our circadian rhythms to regulate many of our bodily functions, most importantly, sleep. Light pollution tricks our brains so that we're unclear of what time of the day it is, even if we're outside. Streetlights shining in through our windows affect our sleep at night making us feel groggy during the day.

As bad as it is for humans, its far, far worse for animals. Many migrating species navigate by the stars. Unable to see the band of the milky way stretching across the sky, animals lose their way, unable to find access to their usual nesting grounds or food sources.

Insects also use the Milky Way as a guide. Artificial light hides the stars. Insects are then attracted to light bulbs, resulting in their demise.

The Great Lakes and central U.S.

Moonlight and Earthlight: The similarities and differences

Moonlight and Earthlight are both forms of reflected sunlight. That is, they both occur when sunlight strikes a planetary body (the Moon or the Earth, respectively) and is reflected off its surface, becoming visible to observers in space or on the surface of another celestial body. Despite this fundamental similarity, there are significant differences between the two.

View from the Gallileo Spacecraft roughly 6.2 million km away

Showing the similarities between Moonlight and Earthlight.

Similarities between Moonlight and Earthlight

Reflected Sunlight:
Both Moonlight and Earthlight are sunlight reflected off the surfaces of the Moon and Earth, respectively.

Phases:
Due to the constantly changing angles between the Sun, Earth, and Moon, both the Earth and the Moon display phases from each other's vantage point. For instance, from the Moon, one would see a full Earth when it's a new moon on Earth. If you were on the Moon when it was full for Earthlings, the dark disk of the "new Earth" would be hard to see.

Varied Intensity:
The intensity of both Moonlight and Earthlight vary based on factors like surface albedo, angle of incidence of sunlight, and phase.

Differences between Moonlight and Earthlight

Albedo:

The Moon has a much lower average albedo (0.12) compared to Earth (0.3). Albedo refers to the reflectivity of a surface; a lower albedo means less light is reflected and more is absorbed. This means that, on average, Earthlight is significantly brighter than Moonlight.

Atmospheric Effects:

The Earth has an atmosphere, which scatters incoming sunlight, causing phenomena like the blue color of the sky and the reddening of sunsets. The Moon lacks such an atmosphere, so moonlight is not subject to atmospheric scattering. This also means that Earthlight can have varying colors based on Earth's atmospheric conditions, while moonlight is typically a bright, white-yellow light.

Illumination of the Dark Side:

During its phases, the dark (nighttime) side of the moon can be faintly lit by Earthlight, also known as "Earthshine". This phenomenon is visible from Earth when the Moon is in its crescent phase. Conversely, the "dark side" of the Earth (the night side) doesn't get similarly illuminated by moonlight due to the Moon's much lower albedo and brightness.

Surface Features:

The Earth's surface features (like clouds, oceans, landforms, and vegetation) can dramatically alter the color and intensity of reflected Earthlight. On the other hand, the Moon, with its relatively uniform, barren, gray landscape, produces a more consistent moonlight.

Earth and the far side of the moon as captured by the DSCOVR Spacecraft

Auroras

The auroras, the mesmerizing dance of lights in Earth's polar skies, invite us to explore the fascinating world of physics that unfolds high above our heads. These radiant phenomena result from interaction between the solar wind and Earth's magnetic field and atmosphere.

But where do the Aurora come from? Views of the Sun though protective solar filters can reveal tiny sunspots, though each spot is frequently larger than the Earth itself! Sunspots are magnetic storms on the Sun. Solar flares, bright flashes of visible light and X-rays, occasionally erupt from sunspots. When the radiation from a flare arrives at Earth 8 minutes later, high-frequency radio signals on Earth can be disrupted, affecting Air Traffic control radio, among other things. After a few minutes, everything returns to normal. Or so we think.

Larger sunspots can eject billions of tons of super-hot plasma into the solar system. From times to time, one of the Coronal Mass Ejections (CME), as they are officially named, is directed towards the Earth. A CME contains so much material it carries with it its own magnetic field!

Aurora Borealis captured from the ISS by Samantha Cristoferetti

A NASA satellite called ACE, which is parked in the Lagrange 1 point between the Earth and Sun, detects the incoming particle stream from the CME and calculates its trajectory and strength. The information is relayed to Earth. If a geomagnetic storm is predicted, air traffic control sends a notification to polar flights to avoid northern latitudes due to increased risk of radiation exposure. Also, electrical utilities prepare for overheating of transformers and induced currents in transmission lines. Satellite launches are delayed to avoid increased atmospheric drag on satellites at low orbital altitudes.

The same high-energy particles that caused all these problems for utilities, airlines, and satellite providers, have another effect. As the particles penetrate Earth's magnetosphere, they become trapped by Earth's magnetic field. They are directed downward through Earth's ionosphere along the magnetic field lines and pump energy into the molecules making up our atmosphere.

Oxygen atoms glow bright green and sometimes red, while nitrogen gives off a purple light.

Only rarely are people living at mid-northern and mid-southern latitudes treated to this other type of Earthlight. But those lucky few who have witnessed, first hand, the Aurora, otherwise known as the Northern Lights, will forever have the spectacle seared into their memories.

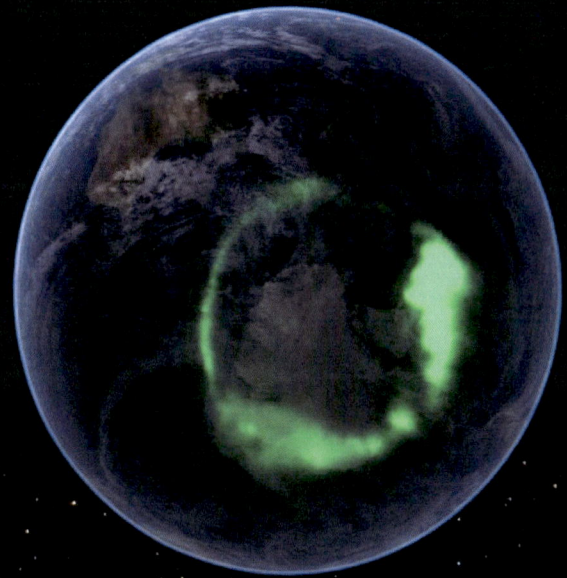

Aurora Australias

Lunar Eclipses

The glow of a thousand sunsets stretches out to the cosmos, surrounding a cone of shadow that we earthlings experience as "the Night". For months on end, the Earth's shadow in space is only detectable when the speeding specks of orbiting satellites wink out while they travel through it.

About twice a year the Earth, Moon, and Sun align and the Moon makes its own passage through Earth's umbra. At that time, only Earth's sunset light, the blue scattered away and only the red remaining, can reach the moon. We call this a blood moon, otherwise known as a lunar eclipse.

The moon goes around the Earth about once a month, so why don't we see lunar eclipses every month? The Moon's orbit is tilted by about 5 degrees away from the plane of our solar system. The Moon's orbit crosses though the plane at two "nodes". Most months, when the Moon is full, it simply misses Earth's shadow, either passing above or below it. But, when the Moon is full while it's near a node, we experience a lunar eclipse somewhere on Earth.

Lunar Eclipse by NASA's Joel Kowsky Composite Lunar Eclipse image by NASA's Norah Moran

Earth Eclipse from the Moon

Here's something you've probably never thought about. What does an eclipse look like from the Moon?

When a spaceship orbiting the Earth passes into Earth's shadow (as it does every 90 minutes or so) we call this "eclipse." The further away from Earth you are, the less frequently Earth eclipses occur.

When we see a total Lunar eclipse on Earth, a resident of the Moon would see a total eclipse of the Sun by the Earth.

Here's the interesting thing. On Earth, the Moon and Sun appear to be the same size. However, from the Moon, the Earth would appear much larger than the Sun!

Eclipse photographed by Apollo 12 during its trans-earth voyage.

Solar Eclipse viewed from the Moon

You might wonder what a solar eclipse, on Earth, would look like from the Moon. Truth is, it would be barely perceptible. The Moon's shadow only makes a small dark patch on the Earth when viewed from Space.

If you think about it, the Moon is only about as wide as North America. During the eclipse, the Moon's shadow forms a cone that stretches toward the Earth. By the time its shadow reaches the Earth, it is only about 200 kilometers wide, covering an area smaller than the state of Vermont.

Earth Eclipse as seen from the Moon (simulated in Stellarium)

Solar Eclipse from Space

The Moon is, on average, 384,400 kilometers from Earth. On the ISS, astronauts are only about 400 kilometers up. From this altitude, the Moon's shadow appears as a dark splotch on Earth's surface.

As the Moon passes between the Earth and the Sun, the shadow sweeps across the Earth's surface. This is not just an extraordinary sight; it's a testament to the intricate physics of shadows.

The shadow consists of two components: the umbra and the penumbra. The umbra is a cone of darkness. Here, for the observer on Earth, the Sun is wholly hidden by the Moon. Within the umbra, day transforms into night. The penumbra, on the other hand, is a broader shadow where the Sun is only partially obscured, in other words, a partial eclipse.

The dance of these shadows is a complex interplay of geometry and distances between the Earth, Moon, and Sun, coupled with their relative sizes. The umbra and penumbra are not mere visual phenomena; they are the visual expression of celestial mechanics.

Solar Eclipse as seen from the ISS

Sian's reflections

I have been obsessed with light my entire life to the point where I even contemplated becoming a physicist. In school, I loved learning about the electromagnetic spectrum and how visible light is just a tiny part of the energy that fuels our universe.

I've researched how early civilizations worshipped the sun and I even incorporate a lot of sunlight metaphors into my artwork. I also love laying in areas in my house where sunlight is piercing through the windows and finding sunbeams illuminating all the dust floating around in the air.

I proudly declare that I am a Sun Devil and not just because I went to Arizona State University, but because I have always loved sunlight. I am equally obsessed with the Moon and Moonlight.

I was born 8 months after Neil Armstrong took those famous first steps on the Moon so I consider myself to be a Moon celebration baby! Sunlight and Moonlight are two words that have always been part of my orbit since my birth and it wasn't until I went to space that I added the word EarthLight.

EarthLight transformed the way I saw our planet and my relationship to both sunlight and moonlight. As a result, it is now my mission to bring EarthLight down to Earth.

Earth from the ISS

Chapter 2

MOONLIGHT REFLECTIONS

The Physics of Lunar Phases

Moonlight Reflections
5 Haiku's

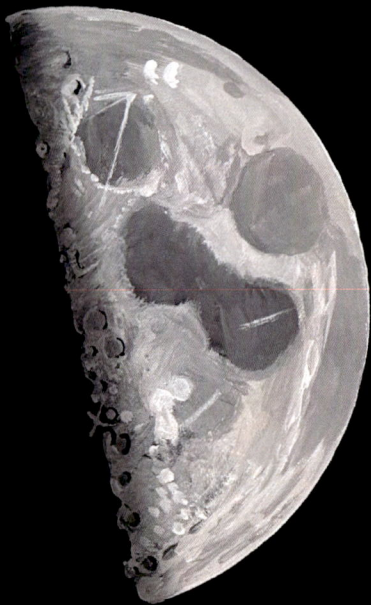

Moon's shimmering light,
Reflects upon the still mind,
Night mirror of dreams.

Moonlight's gentle touch,
Caresses the weary Earth,
Peaceful serenade.

Luna, patient moon,
Witness to Artemis flight,
Women lead the way.

Lunar footprints mark
Artemis' chosen path,
Fulfilling her dream.

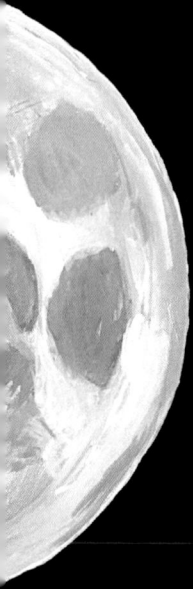

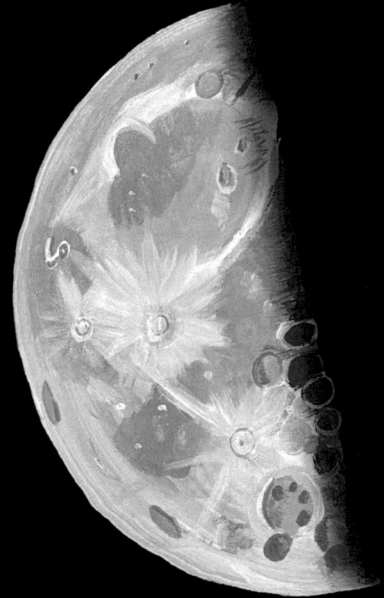

Moon's guiding lantern,
Peeks through the veil of darkness,
Nature's hidden grace.

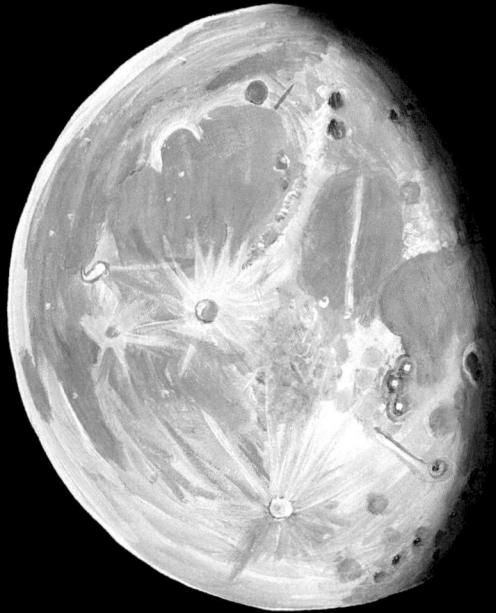

Dr. Sian Proctor

During the full moon phase, the entire side of the Moon facing Earth is illuminated by sunlight.

First

Gibbous Moon

Full Moon

The phases of the Moon, a familiar nighttime sight, are a testament to the intricate ballet of sunlight, Earthlight, and the Moon's orbital dance.

In the new moon phase, the Moon's Earth-facing side is a stranger to sunlight, illuminated primarily by Earthlight. This subtle illumination gives rise to the new moon's ethereal "ashen glow."

As the Moon waltzes around Earth, an increasing fraction of the face we see is lit with sunlight, marked by the transition through the waxing crescent, first quarter, and waxing gibbous phases.

Quarter

During the new moon phase, the side of the Moon facing Earth receives minimal sunlight.

Crescent Moon

With the advent of the full moon phase, sunlight bathes the entire Moon's Earth-facing side.

With an albedo of about 0.12, the full moon at its luminous peak reflects a mere 12% of the striking sunlight. Yet, the full moon's perceived brightness is significantly less than 12% of the Sun's brilliance, primarily due to the Moon's smaller size and greater distance.

Following its luminary climax, the Moon gradually reveals less of its sunlit facade from Earth, commencing the waning phases: waning gibbous, third quarter, and waning crescent. Finally, the Moon reprises its role as the new moon, bringing the lunar cycle full circle.

Young Moon

The Interplay of Earthlight and Moonlight: A Quantitative Perspective

The dance between Earthlight and Moonlight is a fascinating spectacle. Resting about 384,400 km away from Earth, the Moon serves as an immense reflector, bouncing sunlight, altered by the interaction with the Lunar surface back to us, and this spectacle is what we cherish as Moonlight.

384,400 km (238,855 miles)

Earth

Moon

It takes light 1.25 seconds to travel each way.

Because of the Moon's albedo, about 88% of the sunlight that caresses the lunar surface either gets absorbed or scattered in different directions.

During the slender crescent phases, sunlight scarcely touches the Earth-facing side of the Moon. Here, Earthlight, rather than Moonlight, commands the Moon's luminescence when viewed from Earth.

Interestingly, Earthlight on the Moon can outshine Moonlight on Earth, primarily because Earth, being larger and having a higher average albedo, reflects more sunlight. It's a captivating game of cosmic reflections.

> The Earth, with a larger size and higher average albedo, reflects more light than the Moon.

Views from the Galileo Spacecraft

We just unraveled the complex physics of the Sun-Earth-Moon system, zeroing in on their intricate interactions and their profound influence on the Earthlight phenomenon. We'll explore the transformative impact of the "Earthrise" photo on humanity's consciousness and environmentalism, underlining how this iconic image ignited a global epiphany. We also painted a vivid picture of Earth's albedo and its fluctuations, shedding light on its pivotal role in shaping Earthlight and steering climate change. There is an intriguing dynamic between Earthlight and Moonlight, laying bare the importance of the Moon's distance from Earth and its albedo in this cosmic spectacle.

Unveiling the Secrets of the Moon: Why We Only See One Side

An enduring celestial enigma has intrigued humanity for millennia: why do we only see one side of the Moon from Earth? The answer lies in the fascinating physics of tidal locking, also known as synchronous rotation.

The newly-formed Moon rotated rapidly, but the gravitational tug of the much more massive Earth imposed a torque on the Moon's oblong shape, progressively slowing its spin until it became synchronized with its orbit around Earth.

This synchronization means the Moon uses the same period to complete one axial spin and one orbit around Earth, leading to the intriguing result of us always observing the Moon's near side.

Tidal locking, or synchronous rotation, results in the same side of the Moon always facing the Earth, similar to two people holding hands and twirling.

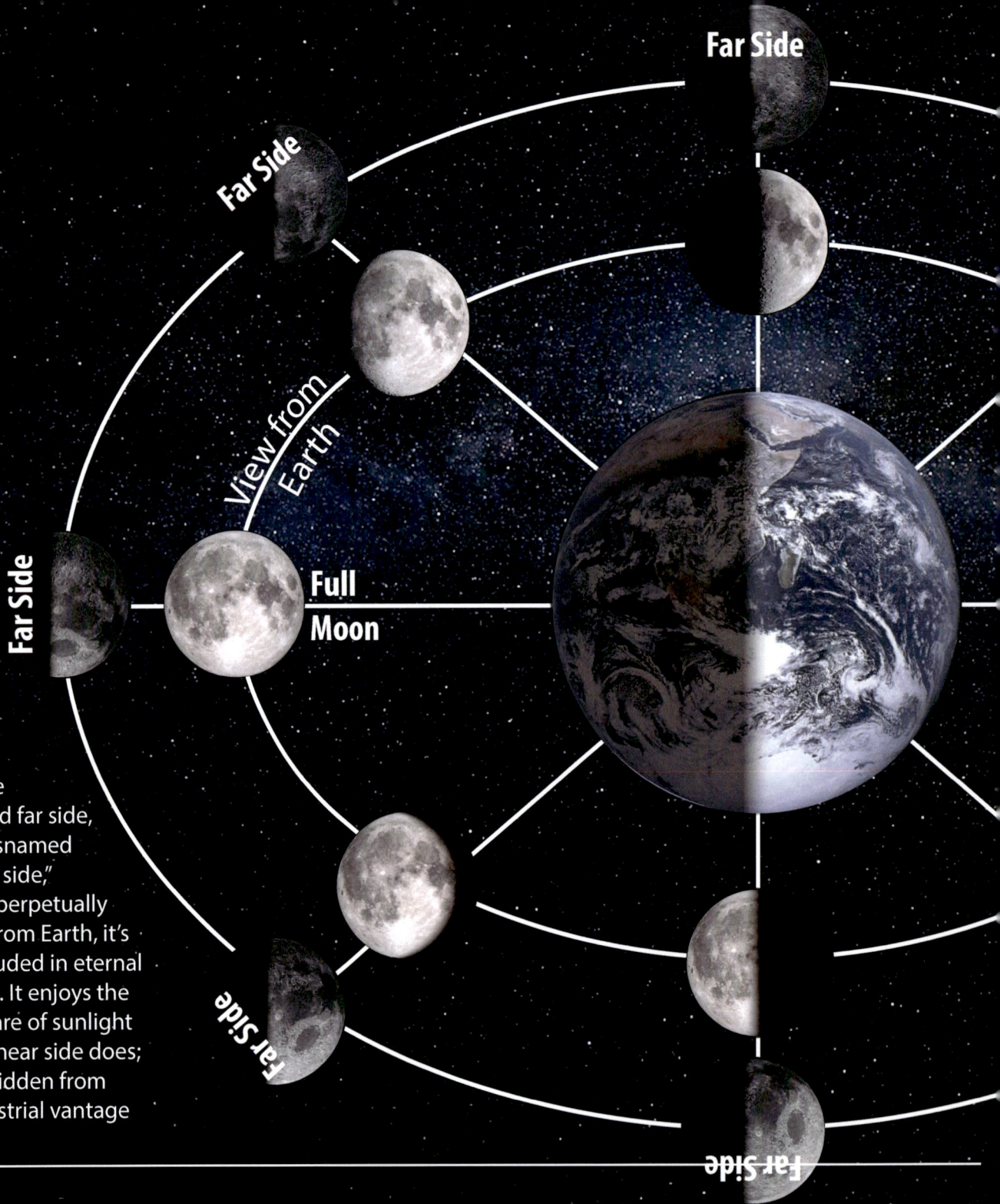

Far Side

Far Side

Far Side

View from Earth

Full Moon

Far Side

Far Side

Far Side

While the concealed far side, often misnamed the "dark side," remains perpetually unseen from Earth, it's not shrouded in eternal darkness. It enjoys the same share of sunlight that the near side does; it's just hidden from our terrestrial vantage point.

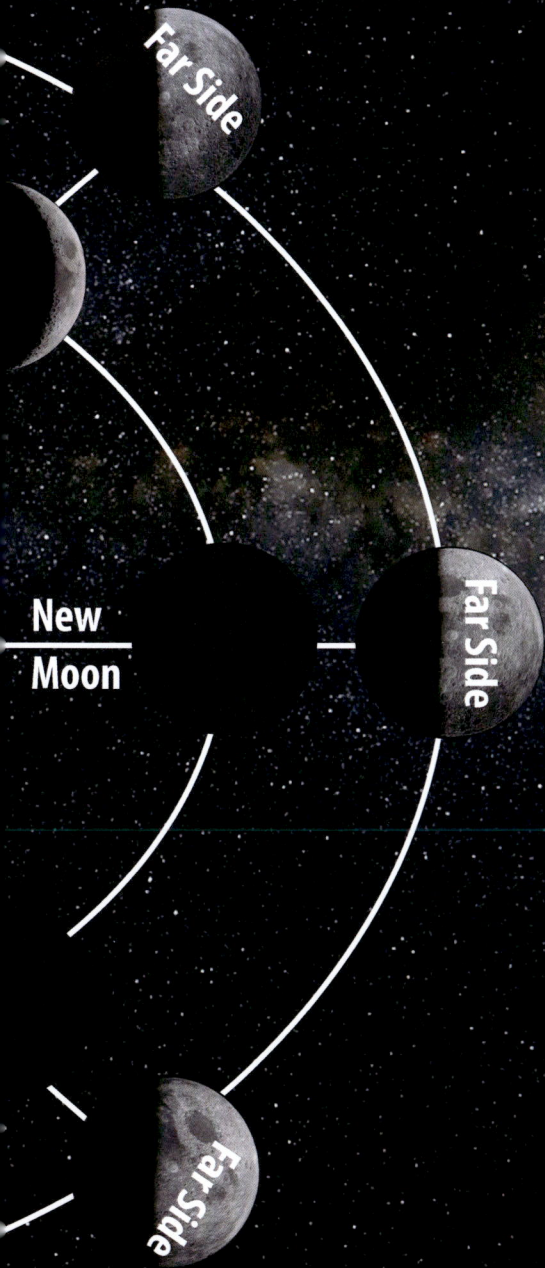

Far Side

Far Side

New
Moon

Far Side

Diagram showing how the far side of the
Moon is illuminated.

The Influence of Moonlight: A Touch on Human Body and Soul

Our nearest celestial companion, the Moon, has been at the heart of innumerable legends and folk tales, many revolving around its sway over the human psyche and body. While Moonlight's physical effects on human physiology are virtually nonexistent, its psychological and emotional imprints are compelling and profound.

From a physical standpoint, Moonlight lacks both the strength and the ultraviolet radiation to influence our bodies like sunlight does. Nonetheless, the Moon and its cyclic phases have been deeply interwoven with human psychology and emotion throughout the annals of history.

Research hints at the possibility of lunar phases subtly influencing human sleep patterns and moods, although the exact workings remain shrouded in mystery.

Here are ten examples of how the Moon and its phases have had a psychological and emotional impact on humans throughout history:

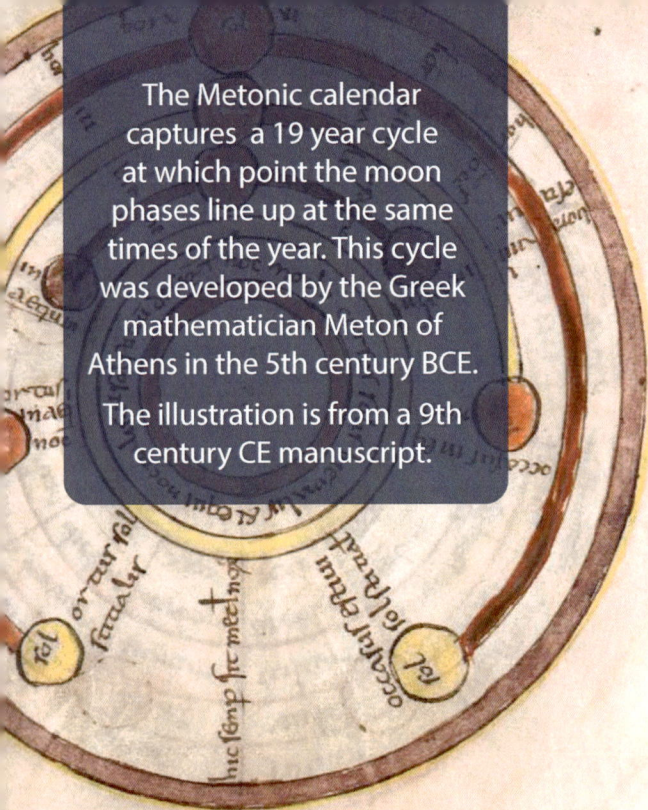

The Metonic calendar captures a 19 year cycle at which point the moon phases line up at the same times of the year. This cycle was developed by the Greek mathematician Meton of Athens in the 5th century BCE.

The illustration is from a 9th century CE manuscript.

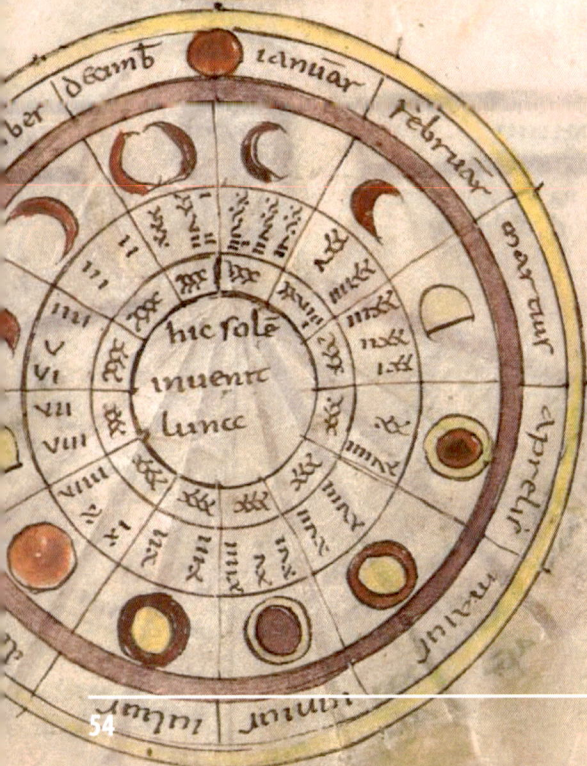

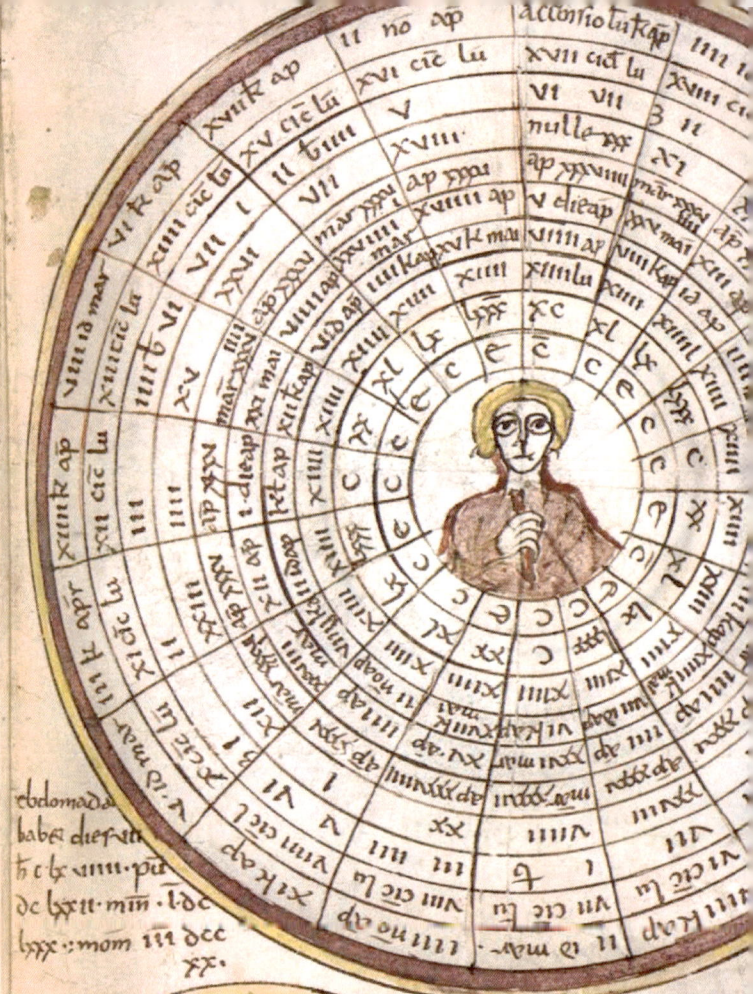

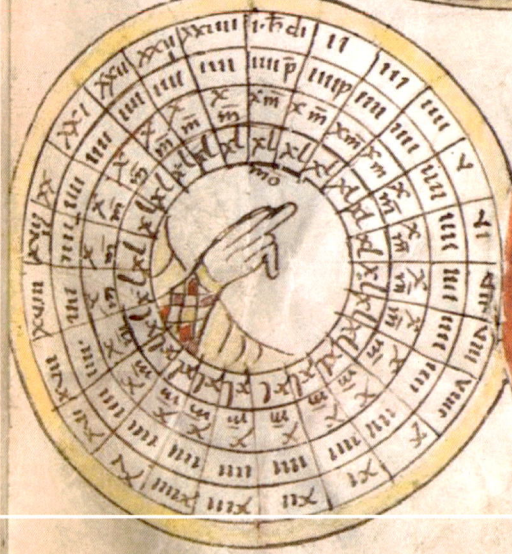

The Coligny Calendar is a bronze plaque inscribed with a calendar following the Metonic cycle. It was created in 2nd century CE Gaul and is the oldest Celtic lunisolar calendar discovered.

1 Lunar Calendars:

Many ancient civilizations, like the Mayans and the Celts, developed lunar calendars, acknowledging the moon's phases for timekeeping. This reverence for lunar cycles suggests a deep psychological and emotional connection with the Moon.

2 Agricultural Scheduling:

Some traditional societies scheduled their agricultural activities according to the Moon's phases, believing the full moon would lead to more abundant harvests. This implies an emotional belief in the Moon's influence on life cycles.

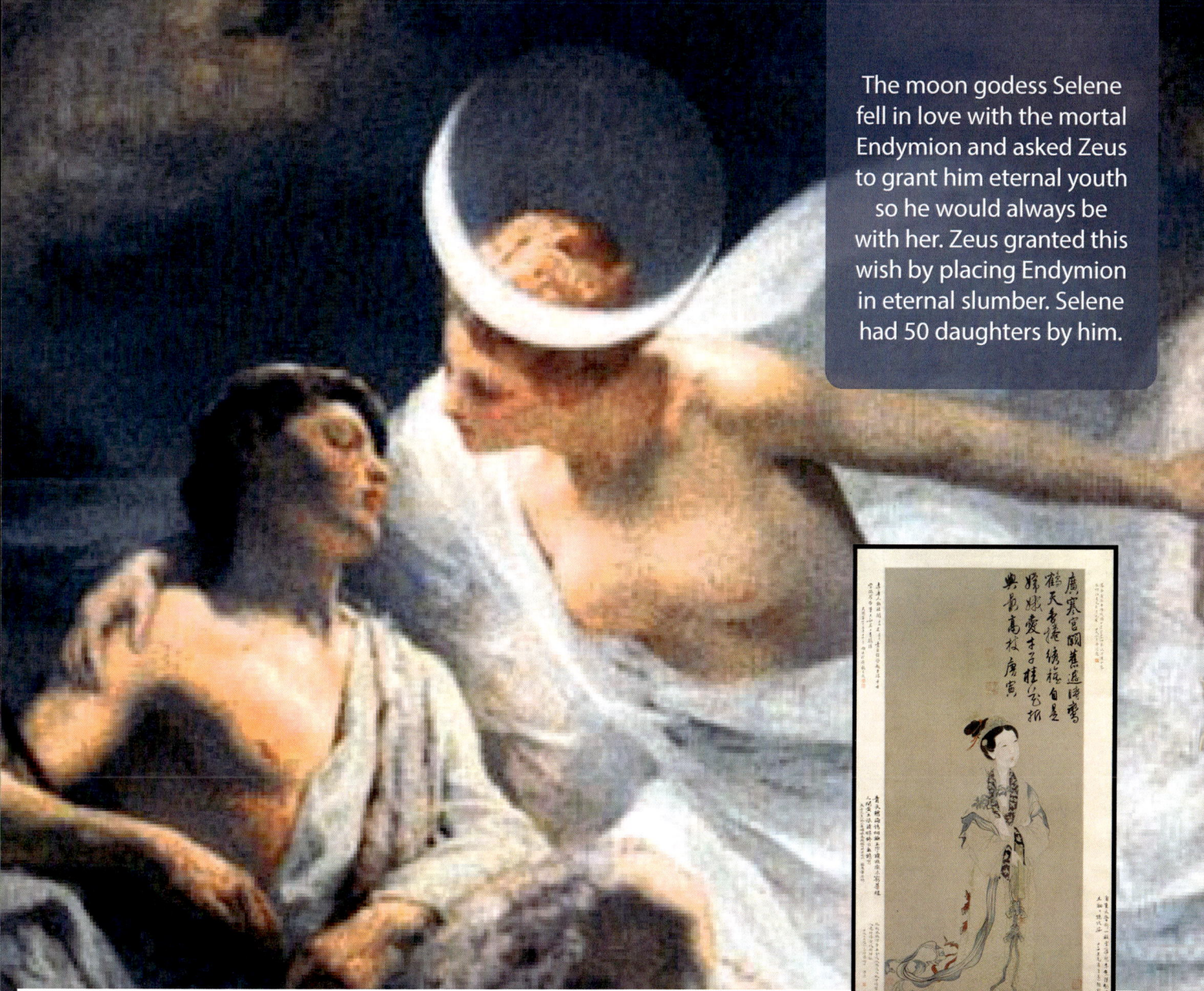

The moon godess Selene fell in love with the mortal Endymion and asked Zeus to grant him eternal youth so he would always be with her. Zeus granted this wish by placing Endymion in eternal slumber. Selene had 50 daughters by him.

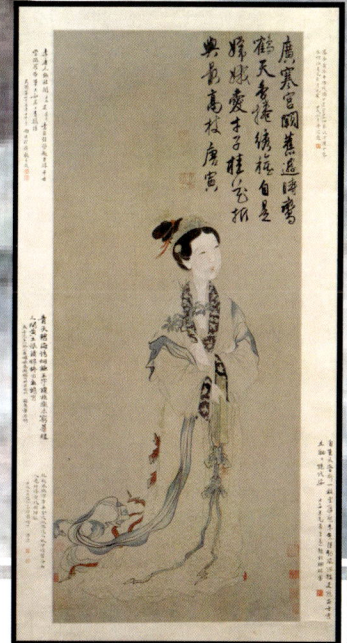

廣寒宮闕舊遊時鸞
鶴天香遶綽枝自是
嫦娥愛才子桂花偏
與月高枝唐寅

3

Moon Deities:

In various cultures, the Moon was personified into deities, like Selene in Greek mythology or Luna in Roman mythology, demonstrating an emotional and spiritual connection to the celestial body.

4   Lunacy and Insomnia:

There's a long-standing belief, evident in the term "lunacy," that the full moon affects mental health and can cause insomnia, although scientific evidence is lacking. This belief reveals the psychological impact the Moon has had on human thought.

5 **Moon Festivals:** Many cultures celebrate moon festivals, such as the Mid-Autumn Festival in China and Tsukimi in Japan, suggesting a communal emotional attachment to the Moon.

THE MOON SPEAKS

I, the moon
would like it known - I
never follow people home. I
simply do not have the time. And
neither do I ever shine. For what you
often see at night is me reflecting solar
light. And I'm not cheese! No, none of
these: no mozzarellas, cheddars, bries, all
you'll find here - if you please - are my
dusty, empty seas. And cows do not
jump over me. Now that is simply
lunacy! You used to come and
visit me. Oh do return,
I'm lonely, see.

James Carter

6 Moon in Literature and Poetry:

The Moon has been a significant subject of literature and poetry, often representing mystery, tranquility, or change, illustrating the Moon's emotional and psychological influence on human creativity.

History as Crescent Moon

The horns
    of a bull
        who was placed
before a mirror at the beginning
        of human time;
            in his fury
at the challenge of his double,
            he has, from
                that time to this,
been throwing himself against
                the mirror until
                by now it is
shivered into millions of pieces -
                here an eye, there
                a hoof or a tuft
of hair; here a small wet shard made
                entirely of tears.
And up there, below the spilt milk of
            the stars, one
            silver splinter -
parenthesis at the close of a long sentence,
        new crescent,
        beside it, red
        asterisk of
Mars

Eleanor Wilner

7

Moon and Romanticism:

The Moon often features in romantic settings or narratives, suggesting an emotional link between the Moon and human notions of love and romance.

8 Werewolf Legends:

The myth of the werewolf transforming under a full moon shows a psychological connection between lunar phases and concepts of change and unrestrained behavior.

9 **Moon in Astrology:**
The Moon is a crucial element in astrology, believed to govern emotions and subconscious habits, showing a psychological and emotional attachment to its phases.

The Triple Goddess of Neopaganism is symbolized by the moon in three phases, representing the Maiden, the Mother, and the Crone.

Also associated with witchcraft which was often women's domain. Like the moon itself, occupying a liminal space where power balances shifted.

10

Moon and Femininity:

The Moon's monthly cycles have been likened to the menstrual cycle in women, symbolizing femininity and fertility in many cultures, demonstrating an emotional and psychological connection between lunar phases and human life processes.

Moonlight has held a commanding presence in human culture and spirituality, igniting inspiration, fueling folklore and mythology, and featuring as a ubiquitous symbol in artistic and literary pursuits. A mere glimpse of the moonlit firmament can stir feelings of peace and awe, underscoring Moonlight's remarkable emotional and mental impact.

Artemis Rising, by Dr. Sian Procter

Sian's reflections

In 1510, Leonardo da Vinci published the Codex Leicester where he illustrates the first documented understanding of EarthLight. Through observations of the Moon, Leonardo realized that light from the Sun reflects off planet Earth and illuminates parts of the Moon that are not in direct sunlight.

This phenomenon, now known as Earthshine, enabled him to see those shadowed parts of the Moon and create a sketch in his Codex Leicester.

Fast forward to 2021, and my Inspiration4 flight. As a geoscientist, I know the Earth has a high reflectivity but it wasn't until I was literally floating in the SpaceX Dragon Cupola and being bathed in EarthLight that I truly understood this phenomenon.

It is fitting that my crew members gave me the call sign Leo as homage to being a modern-day Leonardo da Vinci. Someone who combines both art and science. This book's cover photo of me in the cupola being bathed in EarthLight with Leo on my shirt is my favorite moment from space.

The Earth's Silent Glow

In the changing space between day and night

The sun, our fusion power to the cosmos

Ignites the colors of Earth's ethos

A smile and wave to begin the day

With each sunrise, Earth awakes

Bathed in golden hues, a life-giving nectar

Its existence marked amidst a celestial parade

From dawn's embrace to twilight's fall

Earth keeps pace to sunlight's relentless beckoning

Moonbeams cast the night a gentle glow

Physics universal choreography on auto-play

Everything moves and sways to this celestial ballet

Earth's splendor is a ring tone

A reflection echoed across the murky way

We bask in the glory, waiting for an answer

By Dr. Sian Proctor

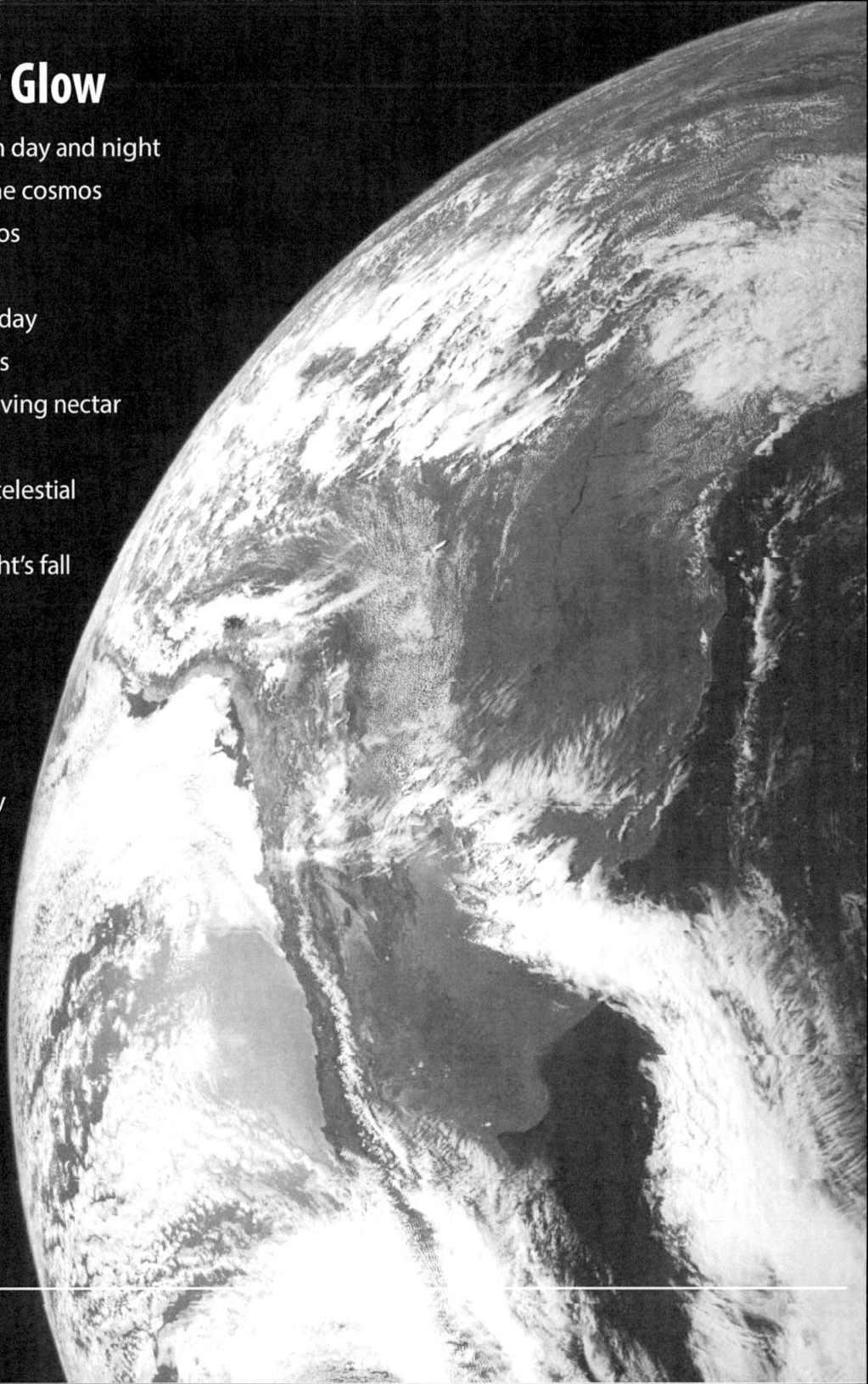

Chapter 3

THE EARTH'S SILENT GLOW

Earthlight Revealed

Observing Earth from Space: A Historical Perspective

In 1957 and 1958, humankind embarked on an unprecedented journey with the launch of the Sputnik from the Soviet Union and the Explorer 1 satellite from the USA. These groundbreaking moments launched the Space Age, offering us a novel perspective on our own planet.

With the launch of the TIROS-1 (Television Infrared Observation Satellite) in 1960, a new age dawned in meteorology, with satellite imagery becoming an indispensable tool for weather prediction. The launch of Landsat 1 in 1972 further revolutionized our relationship with space, enabling us to monitor our environment and manage resources like never before.

Explorer 1 satellite in 1958 marked the beginning of the Space Age for America. Early satellites opened up a whole new perspective of our planet.

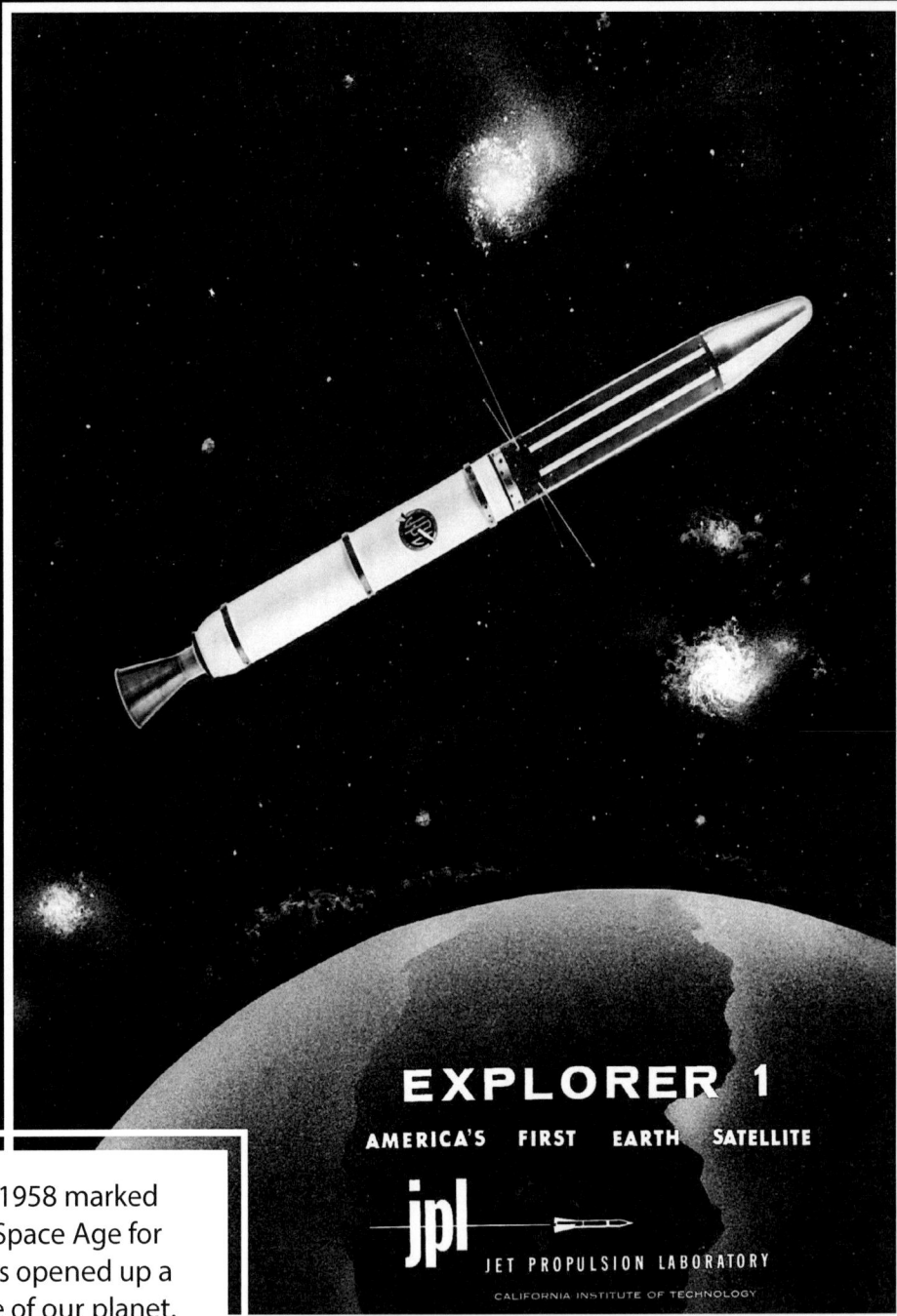

EXPLORER 1

AMERICA'S FIRST EARTH SATELLITE

jpl

JET PROPULSION LABORATORY

CALIFORNIA INSTITUTE OF TECHNOLOGY

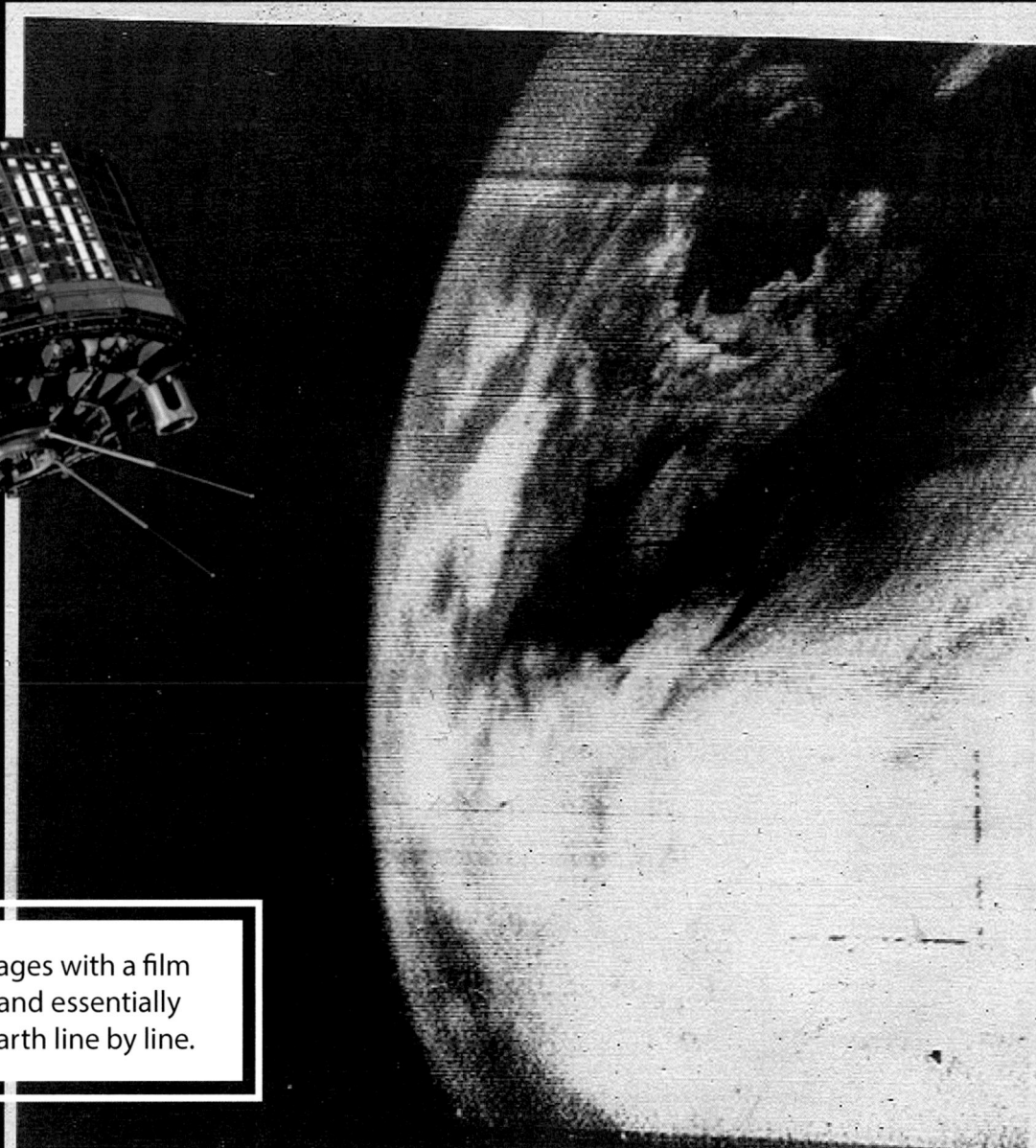

FIRST TELEVISION PICTURE FROM SPACE
TIROS I SATELLITE   APRIL 1, 1960

TIROS-1

TIROS-1 captured images with a film style movie camera and essentially faxed them back to Earth line by line.

In 1972, the launch of Landsat 1 heralded a new era of Earth observation for environmental monitoring and resource management.

Landsat 1 image of Los Angeles and surrouding areas

The 1968 "Earthrise" photo by NASA astronaut Bill Anders shows Earth as viewed from lunar orbit. This image, depicting our planet as a fragile blue marble in the vastness of space, had a profound effect on our collective consciousness. Nothing prepared us for the transformation that would occur in 1968 when the Apollo 8 mission brought us the iconic "Earthrise" image.

National Aeronautics and Space Administration

In the cosmic depths
we discover ourselves

EARTH DAY

www.nasa.gov

Earthrise: A Catalyst for Environmental Awareness

The "Earthrise" photograph is more than a mere picture—it's a testament to a groundbreaking moment in human history. It offered us a startling glimpse of our world as seen from a different celestial body, underscoring Earth's fragility and solitude in the expanse of space. This iconic image acted as a catalyst for global consciousness. It underscored our shared custody of this enchanting blue marble and ignited an unprecedented wave of environmental cognizance. The "Earthrise" photograph became a beacon for the budding environmental movement, inspiring individuals worldwide to champion Earth's preservation and protection.

EARTH DAY - APRIL 22
Union Square
Fifth Avenue

Environmental Action Coalition 235 East 49th St. New York 10017 Telephone 986-9600

The New York Times

LATE CITY EDITION
Weather: Cloudy, rain likely today and tonight. Partly sunny tomorrow.
Temp. range: today 64-50; Wed. 67-46. Full U.S. report on Page 73.

VOL. CXIX... No. 40,997   © 1970 The New York Times Company.   NEW YORK, THURSDAY, APRIL 23, 1970   10 CENTS

PRICES CLIMB 0.4% BUT RATE OF RISE APPEARS TO SLOW

Gain in the Consumer Index for March Lags Behind the 4 Previous Months

INCREASE IS STEEP HERE

Medical Care and Mortgage Interest Are the Major Elements in Changes

By EDWIN L. DALE Jr.
Special to The New York Times
WASHINGTON, April 22— Consumer prices rose strongly again in March but there were signs that the pace of inflation was slowing, the Labor Department reported today.

After adjustment of the data to reflect normal seasonal changes, the Consumer Price Index rose four-tenths of 1 per cent in March, less than the five-tenths rise in February and the six-tenths rate of the three months before that.

What is more, one-quarter of the entire March increase in the index was accounted for by a somewhat artificial recording of an increase in Veterans Administration mortgage interest rates. The ceiling on these rates was recently raised by the Government, but the cost to the borrower was not increased by as much as the face amount of the interest.
Finds Easing of Rate

Millions Join Earth Day Observances Across the Nation

Throngs jamming Fifth Avenue yesterday in response to a call for the regeneration of a polluted environment. View is north from 43d Street, with Central Park in background.
The New York Times (by Patrick A. Burns)

CAMBODIAN CRISIS GROWS AS TROOPS SEEM TO FALTER

With Reds Near, Pnompenh Is Gloomy Over Limited Response to Aid Pleas

CIVIL AVIATION CURBED

Aide Declines to State How Long Army Can Hold Out With the Arms It Has

By HENRY KAMM
Special to The New York Times
PNOMPENH, Cambodia, April 22—An atmosphere of heightening national emergency is overtaking Cambodia.

It is due to evidence that the Cambodian Army is unable to turn back the Vietnamese Communist forces, which at one point are within 15 miles of the capital, and to the limited response to Premier Lon Nol's appeal to all nations for arms aid.

The military authorities closed the Pnompenh airport this evening to all civilian traffic. According to military sources, it is to remain closed until tomorrow in connection with an important military operation. The operation is presumably intended to dislodge the North Vietnamese and Vietcong forces from the district capital of Saang, about 15 miles south of here.
An Appeal to Newsmen
The discouragement over the

U.S. CONCERN SUED WITH 2 IN JAPAN

Westinghouse and Mitsubishi

Mood Is Joyful as City Gives Its Support

By JOSEPH LELYVELD
Huge, light-hearted throngs

Activity Ranges From Oratory to Legislation

By GLADWIN HILL
Earth Day, the first mass

BACKERS OF ROJAS THREATEN REVOLT

But Colombian Government

U.S. Plane Flies In Arms As Trinidad Fights Mutiny

By TAD SZULC
Special to The New York Times
WASHINGTON, April 22—The United States flew a

Moreover, the "Earthrise" image rippled across various domains, from arts and culture to politics and environmental policies. It was instrumental in sparking the first Earth Day in 1970, serving as a poignant reminder of Earth's vulnerability and our shared responsibility to safeguard it. This powerful symbol continues to remind us of our duty to our home planet.

The evolution of satellite technology since then has been nothing short of remarkable. Today, a fleet of satellites orbits Earth, offering a near-constant

The "Earthrise" photo was not scheduled for the mission. When the Apollo 8 crew saw the Earth rising over the lunar horizon, Bill Anders quickly replaced the black-and-white film in his camera with custom Ektachome film developed by Kodak.

Since then, satellite technology has advanced exponentially, offering increasingly detailed and frequent observations of Earth.

Aqua Satellite Monitoring atmosphere and water cycle

Terra Satellite Monitoring pollution

Each Earth observation satellite carries several instruments that use a variety of methods to help scientists learn more about our world and how we can work to protect it. Some of the instruments use "active" instruments that emit electromagnetic radiation, primarily at radio wavelengths. Others are "passive" and collect the information as it comes to them. Observing the Earth from afar is called "Remote Sensing".

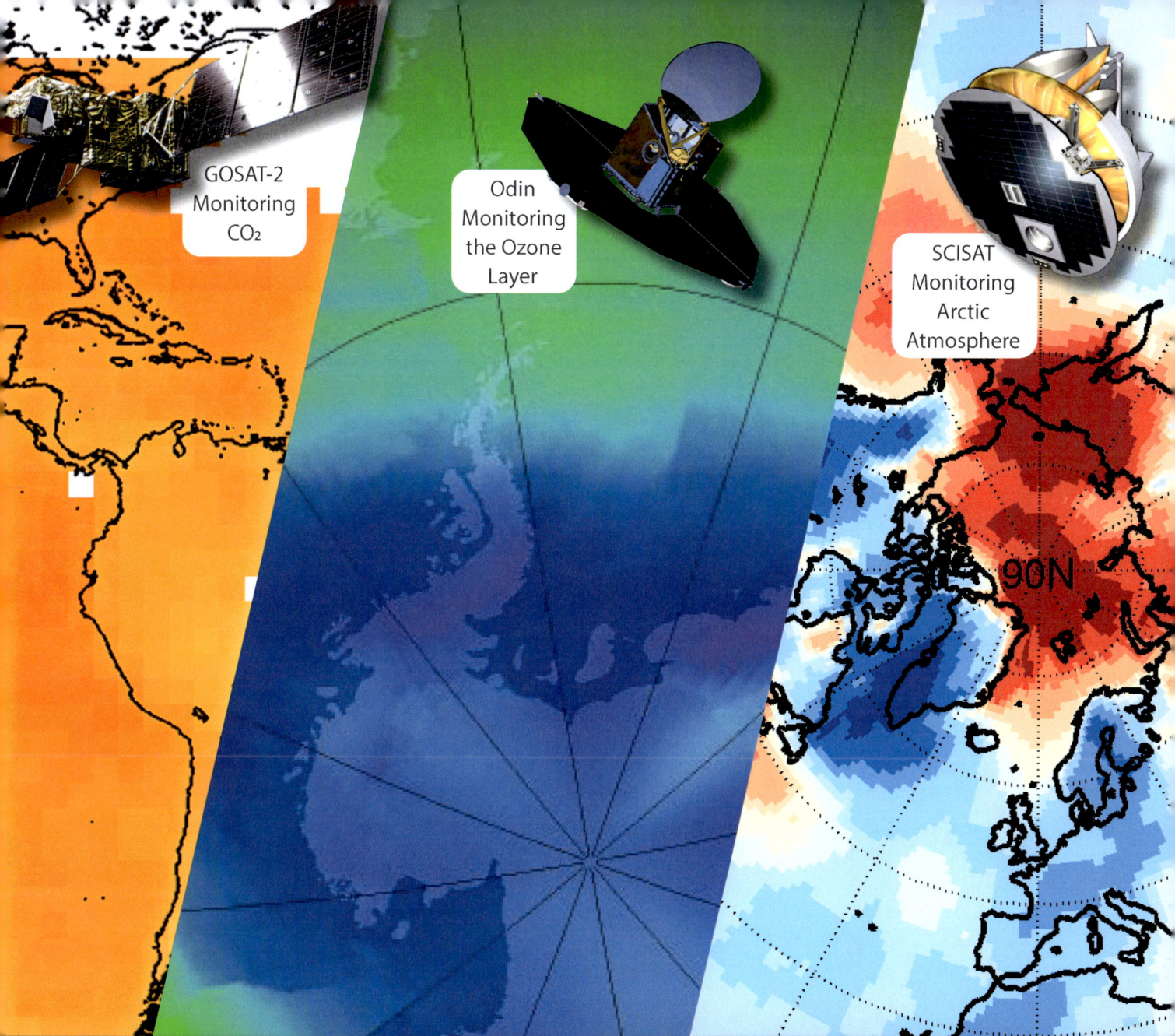

GOSAT-2
Monitoring
CO_2

Odin
Monitoring
the Ozone
Layer

SCISAT
Monitoring
Arctic
Atmosphere

90N

Many different satellites carry identical instruments. This helps to collect more data over more area and allows scientists to compare and verify the information gathered by the different spacecraft. It is the materials that make up the Earth (whether in the atmosphere, ocean, or land) that change Sunlight into Earthlight. We use satellite technology to monitor these changes.

Earth's Albedo and its Variations: An Insight into Earthlight

The concept of albedo—quantifying the proportion of sunlight reflected back into space—plays a starring role in the theater of Earthlight. Known as planetary albedo in the context of Earth, it influences the nature and intensity of Earthlight.

Earth has different albedos on different surfaces. The pristine, snow-covered terrains, with their high albedo, reflect the majority of sunlight. In contrast, the vast ocean surfaces, having a lower albedo, drink in more light.

It is this beautiful array of reflectivity that orchestrates the variations in Earthlight. Not just a mere player in Earthlight's theatre, Earth's albedo also choreographs the dance of our planet's climate.

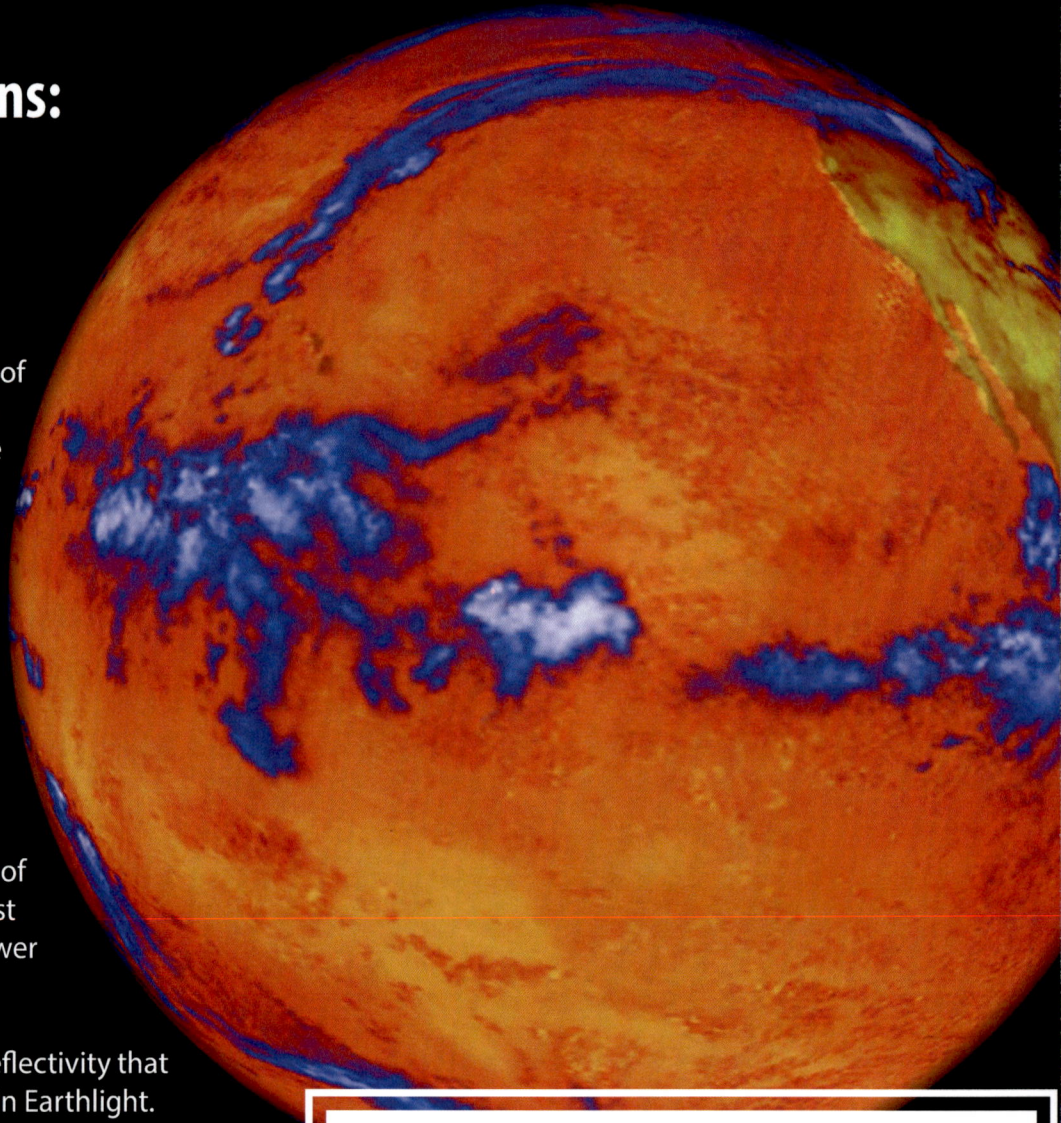

This image shows how much light in reflected (red) vs how much is absorbed (red). A decrease in Earth's albedo, say, due to melting ice caps, can lead to more absorption of solar energy, amplifying global warming.

Different surfaces on Earth have different albedos. For instance, snow-covered surfaces have a high albedo and reflect most of the sunlight, while forests have a lower albedo and absorb more light.

The images below show how changing surface conditions affect Earth's Albedo throughout the year.

Winter

Spring

Summer

Fall

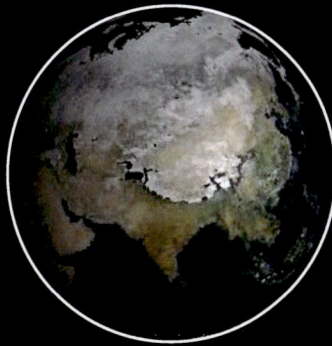

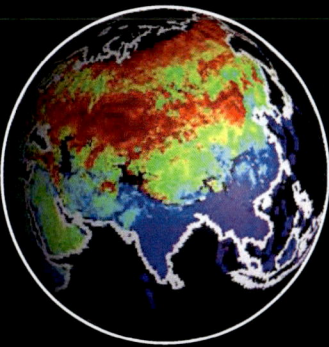
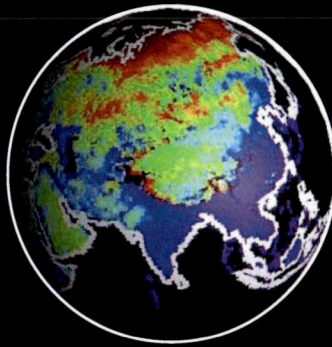
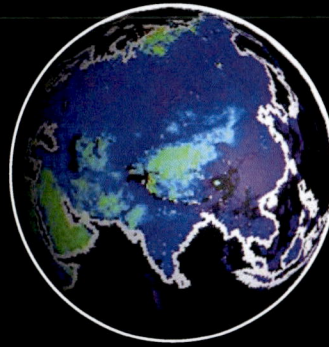
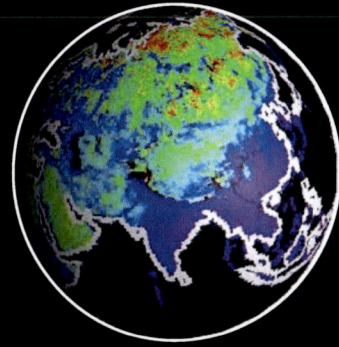

Sian's reflections

Finding satellites streaking across the night sky was one of my favorite things to do as a kid. I would sneak out of my bedroom window late at night and lay out on the roof staring up at the sky for hours. But with all my childhood wonder, I never thought what it would be like to be in space and look back at the Earth.

The notion of seeing our planet from that orbital perspective was still foreign to me until years later when I went to graduate school and became a geologist. The idea of not just looking out into space but being in space and looking back at Earth became the real goal as I began to chase space.

Becoming a geologist also helped me fully understood the link between that small dot of light crossing the night sky and the science behind observing our planet from space.

With my Inspiration4 Mission we were at an altitude of 585 kilometers, which is well above the International Space Station and even the Hubble Telescope. From the orbital vantage point of our SpaceX Dragon capsule, I actually saw satellites from space moving against the backdrop of a darkened Earth at night. It is a memory I cherish because it brought me back to my childhood and those moments on my roof.

I realized that I was now the satellite in space and I was using my own eyes to spy on the Earth and what I discovered was EarthLight.

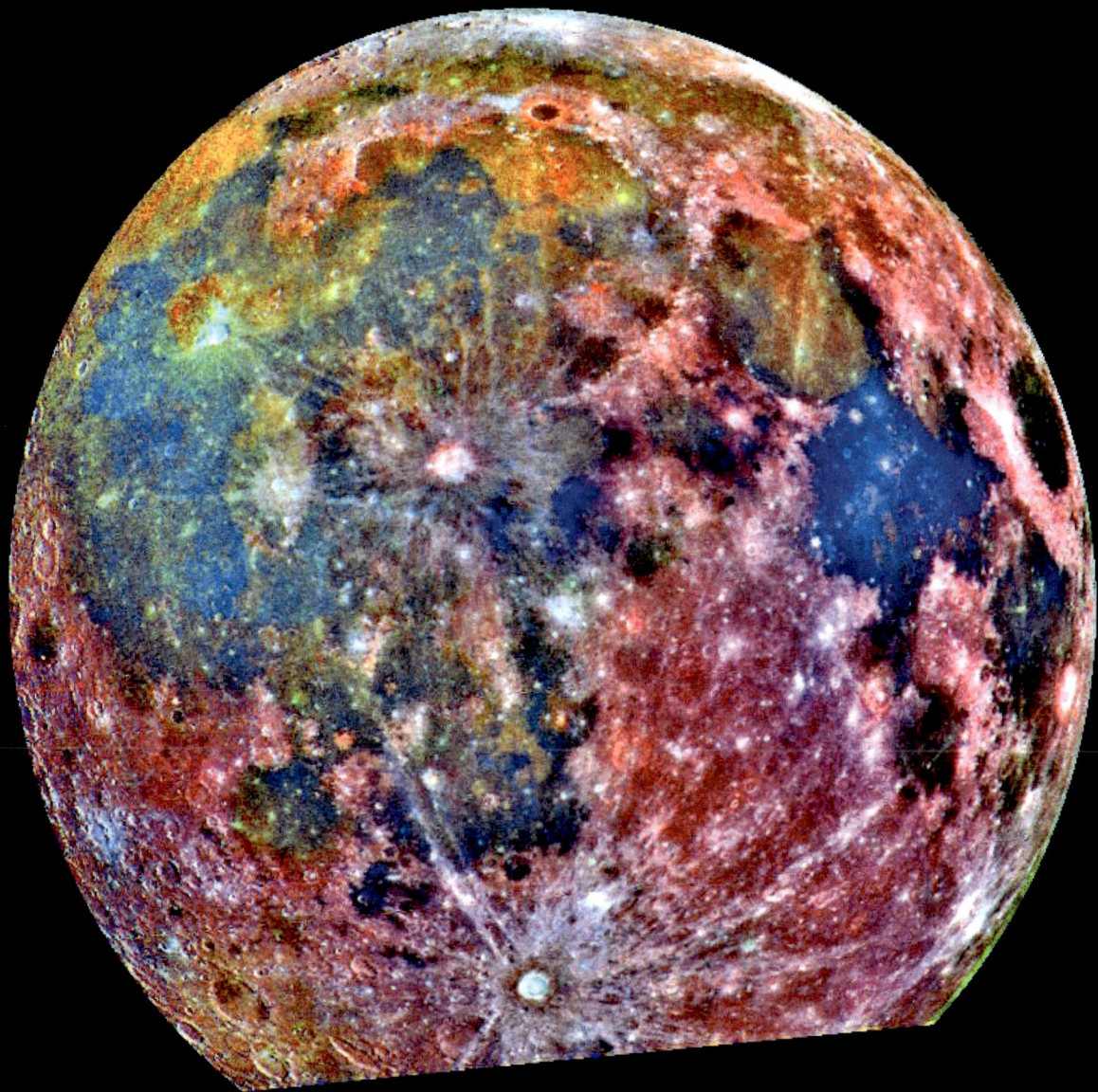

False color image of the Moon showing soil composition

Shrinking of the Aral Sea: The Aral Sea, located between Kazakhstan and Uzbekistan used to be one of the largest lakes on Earth. Once 68,000 square kilometers in size, the lake is now almost completely gone. It is considered one of the most devastating ecological disasters of our time. (Left image, 2014, Right image, 2020)

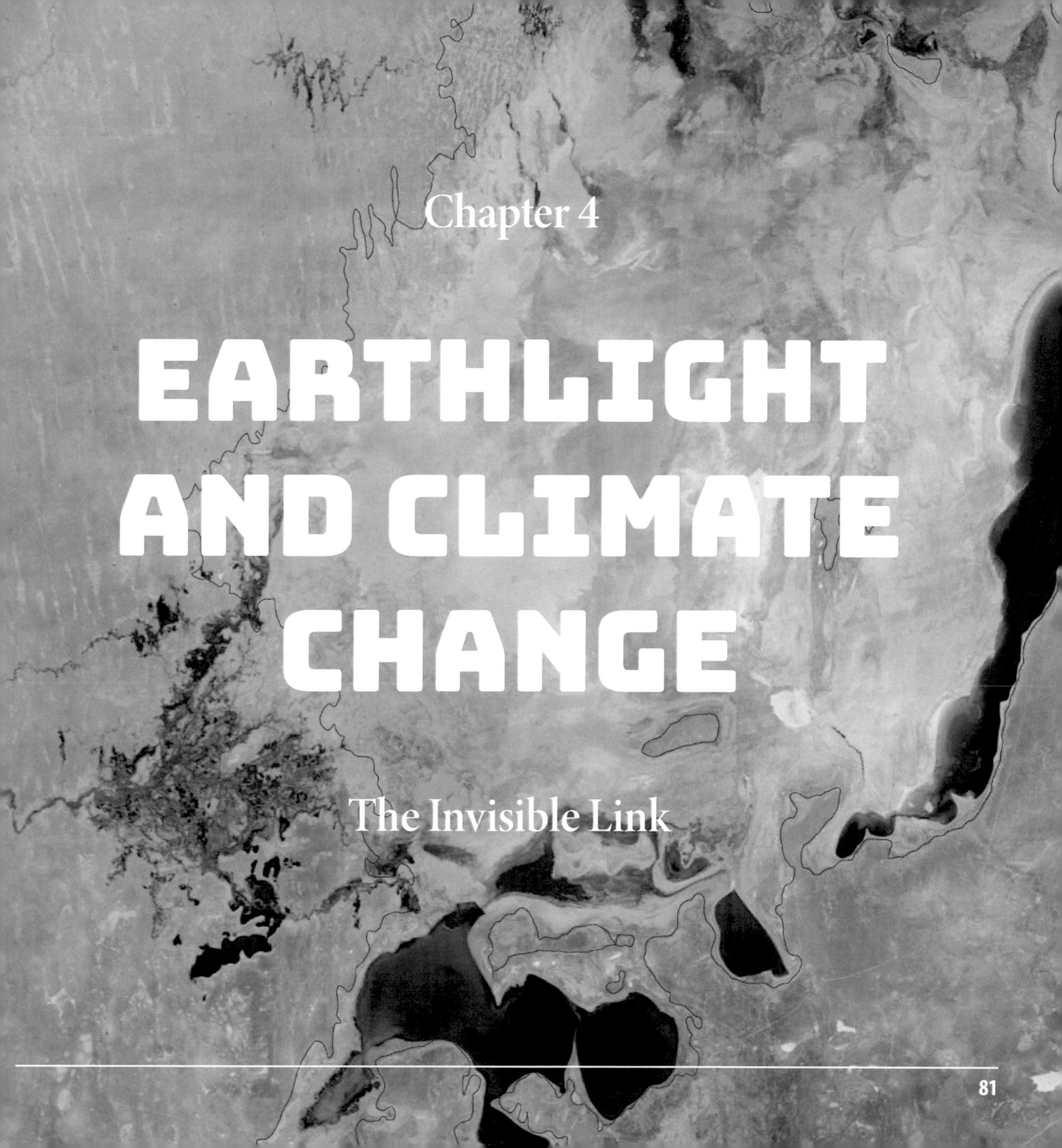

Chapter 4

EARTHLIGHT AND CLIMATE CHANGE

The Invisible Link

EarthLight and Climate Change

From afar you see the Earth glow,
A calling beacon to watch the show.

Concrete constellations stitched across the
terraform,
A testament to the presence of the human
swarm.

Streetlights twinkle upon carbon
creations,
Guiding weary travelers to their
beloved destinations.

Skyscraping towers to celestial heights,
Provide the artificial stars on urban
nights.

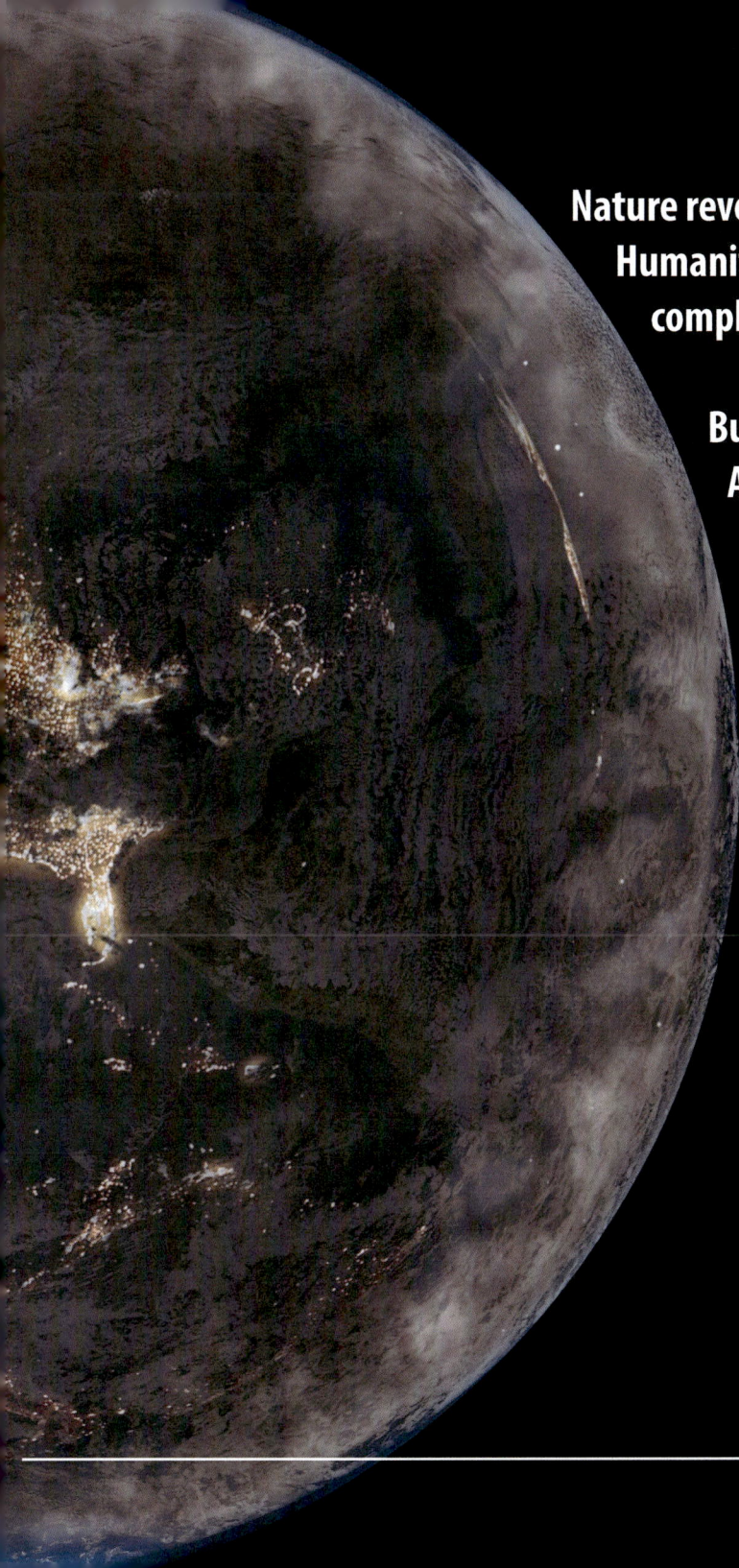

Nature reverberates to the buzz of the beat,
Humanity's thumping imagination is never
complete.

Building a monument without resistance,
A fossil index to the creator's existence.

Tantalizing to watch the evolving
stream,
Earth's radiant beauty shattered by
human dreams.

A mosaic of secrets crashing the shore,
Earth's light-filled amusement,
nevermore.

Dr. Sian Proctor

Earthlight is a crucial resource for climate modeling.

Earthlight, the radiance reflected off our planet's surface, is more than an astronaut's view. It is intimately connected to our climate system's well-being. At the heart of this relationship lies 'albedo' – the reflectivity of Earth's surface.

It can illuminate crucial aspects of our intricate climate system, refining our predictions of climate change and our ability to mitigate its effects. . Changes in albedo, influenced by shifting ice coverage, vegetation patterns, and urbanization, alter Earth's temperature. These changes are mirrored in the quality of Earthlight.

Albedo is a temperature regulator for our planet. High albedo surfaces, such as snow and ice, reflect significant sunlight back into space, offering a cooling effect. But climate change threatens this balance.

As ice melts and albedo diminishes, more sunlight is absorbed, accelerating global warming. Here are real-world examples of the relationship between albedo and climate change.

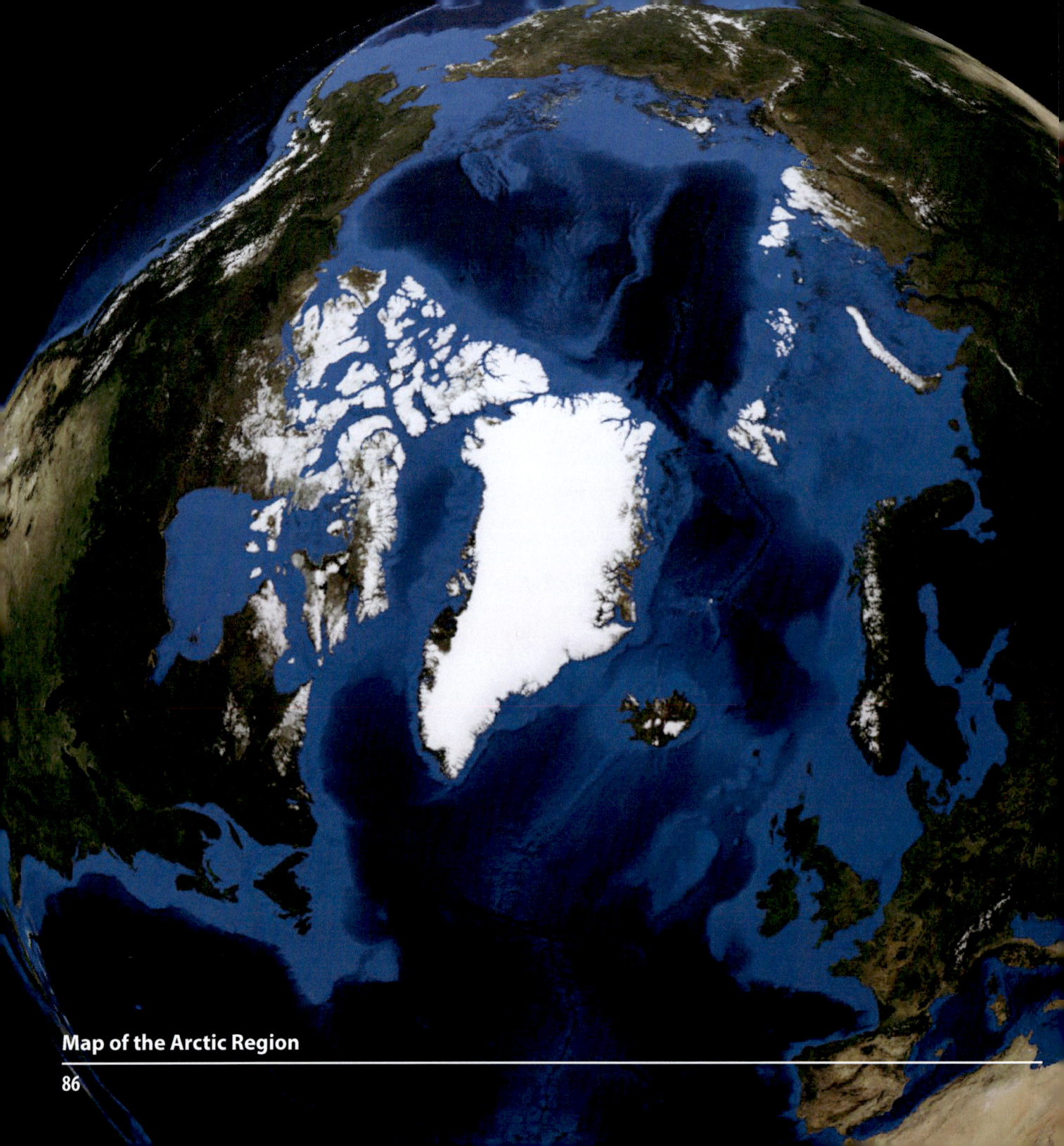

Map of the Arctic Region

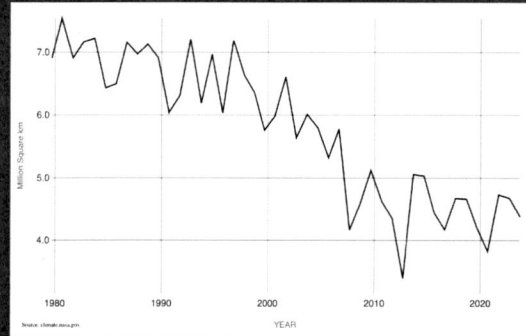

Arctic Ice Melt: According to the National Snow and Ice Data Center (NSIDC), the Arctic has been losing sea ice at a rate of 12.2% per decade. This decrease in white, reflective ice surface has led to a decrease in albedo since more dark ocean water, which absorbs sunlight, is being exposed.

http://nsidc.org

Operation IceBridge examined pools of melt water over the Arctic Sea

Greenland's Ice Sheet: A study published by the European Geosciences Union revealed that the albedo of the Greenland Ice Sheet will decrease by 0.08 by the year 2100. This might sound insignificant, but it means the ice sheet is absorbing far more solar energy, which has a major affect on the climate. https://doi.org/10.5194/tc-10-477-2016

Apusiaajik Glacier, Greenland

Apusiaajik Glacier, Greenland

Antarctic Peninsula: British Antarctic Survey reported a reduction of about 15% per decade in the albedo of the Antarctic Peninsula glaciers since the late 20th century. This change, primarily due to ice melt and decreased snowfall, contributes to global warming.

Larsen Ice Shelf in Antarctica

SOUTH
SHETLAND
ISLANDS

Weddell
Sea

Palmer Land

Antarctic
Peninsula

Bellingshausen
Sea

Ellsworth
Land

Vins
Mas

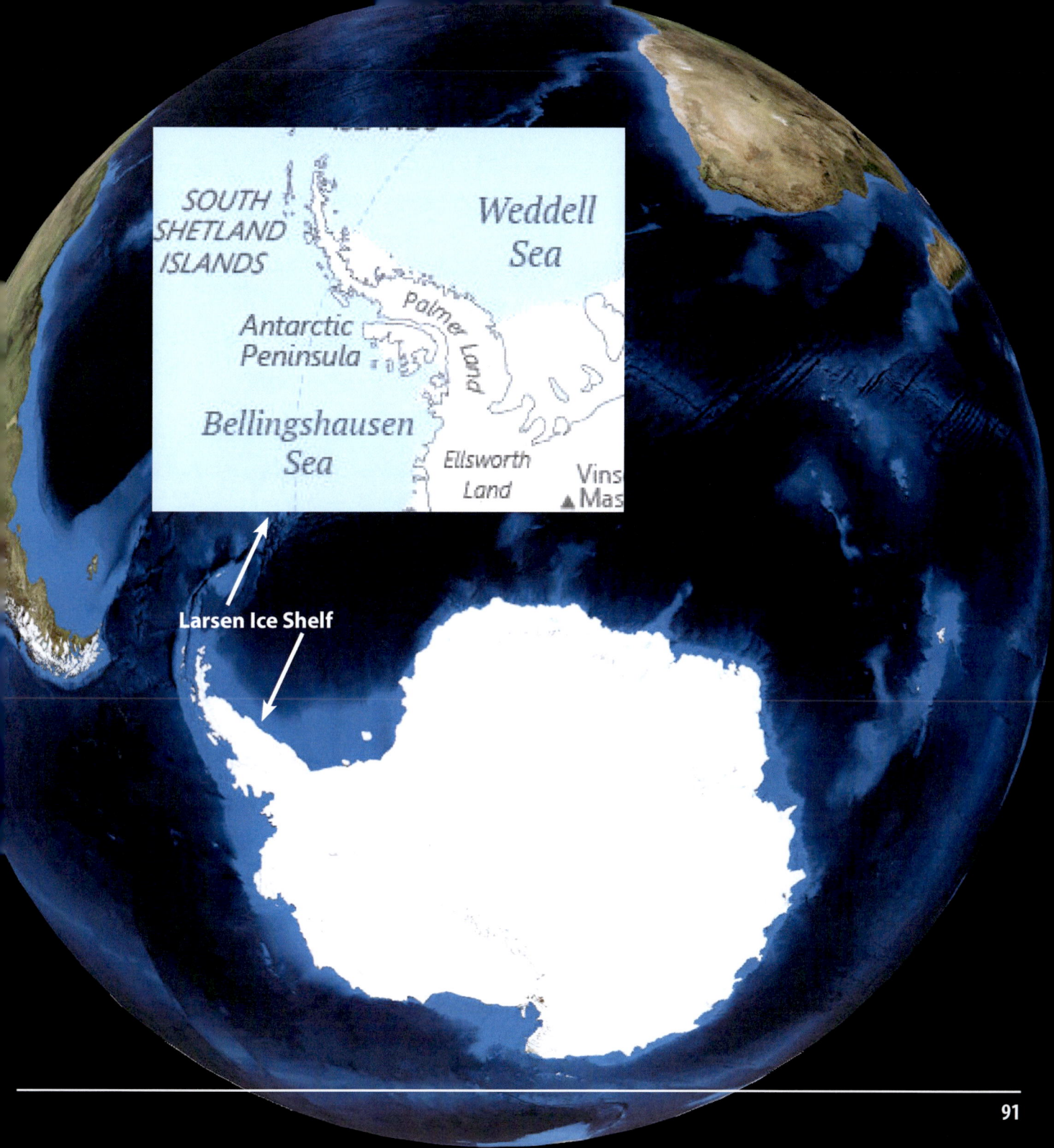

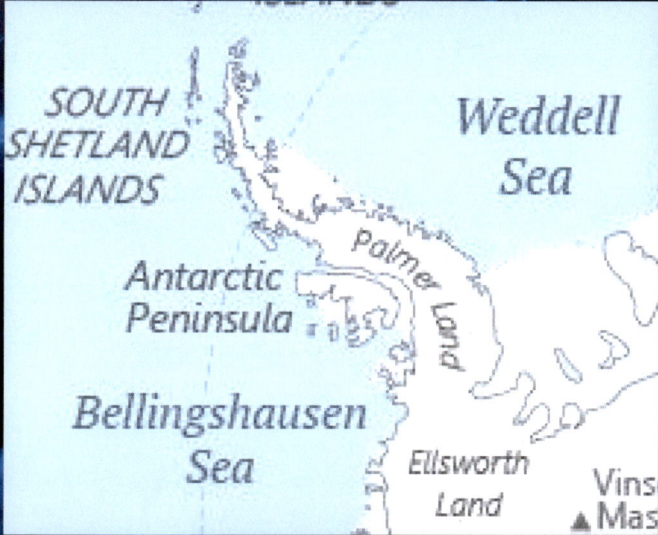

Larsen Ice Shelf

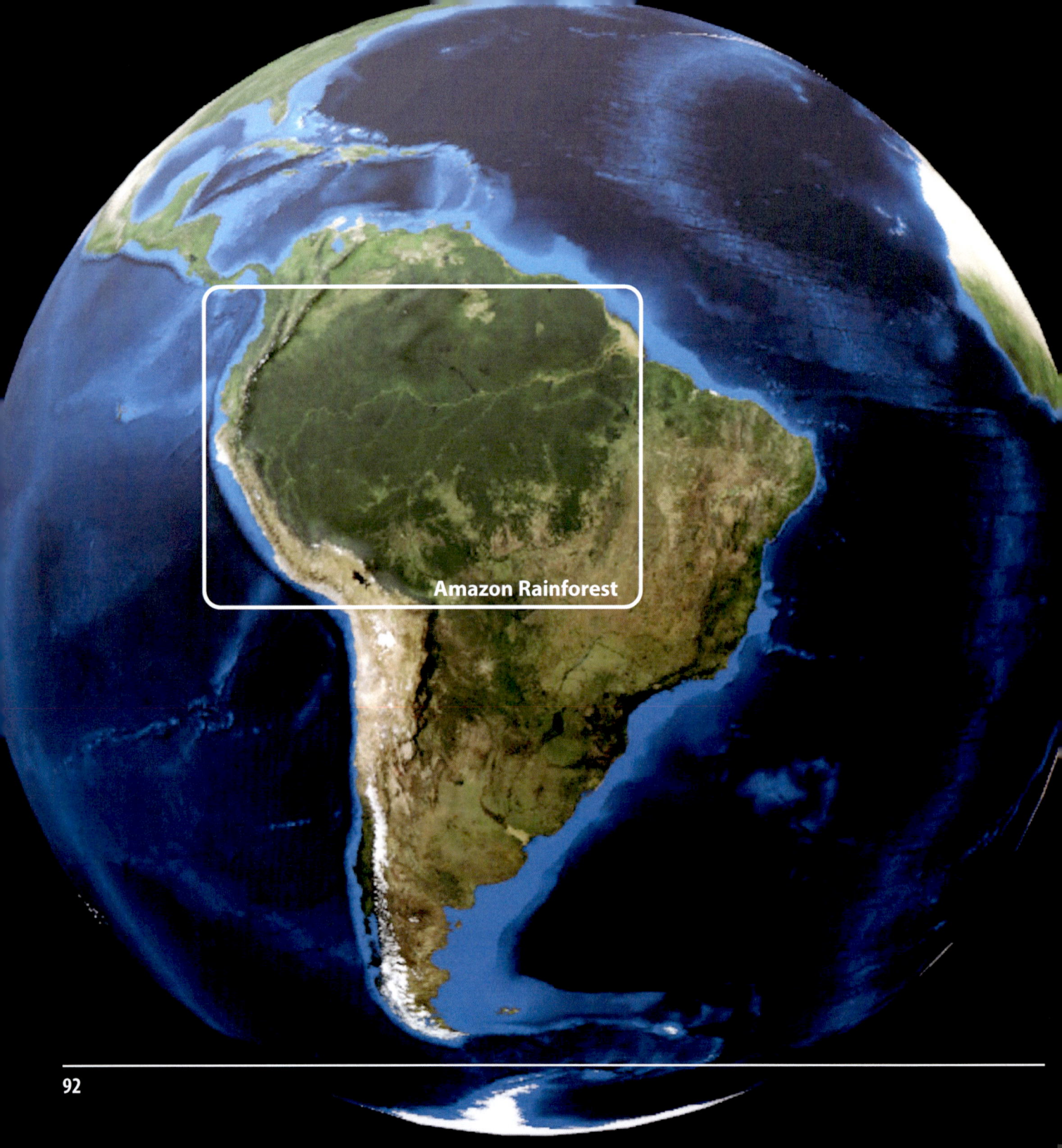

Amazon Rainforest

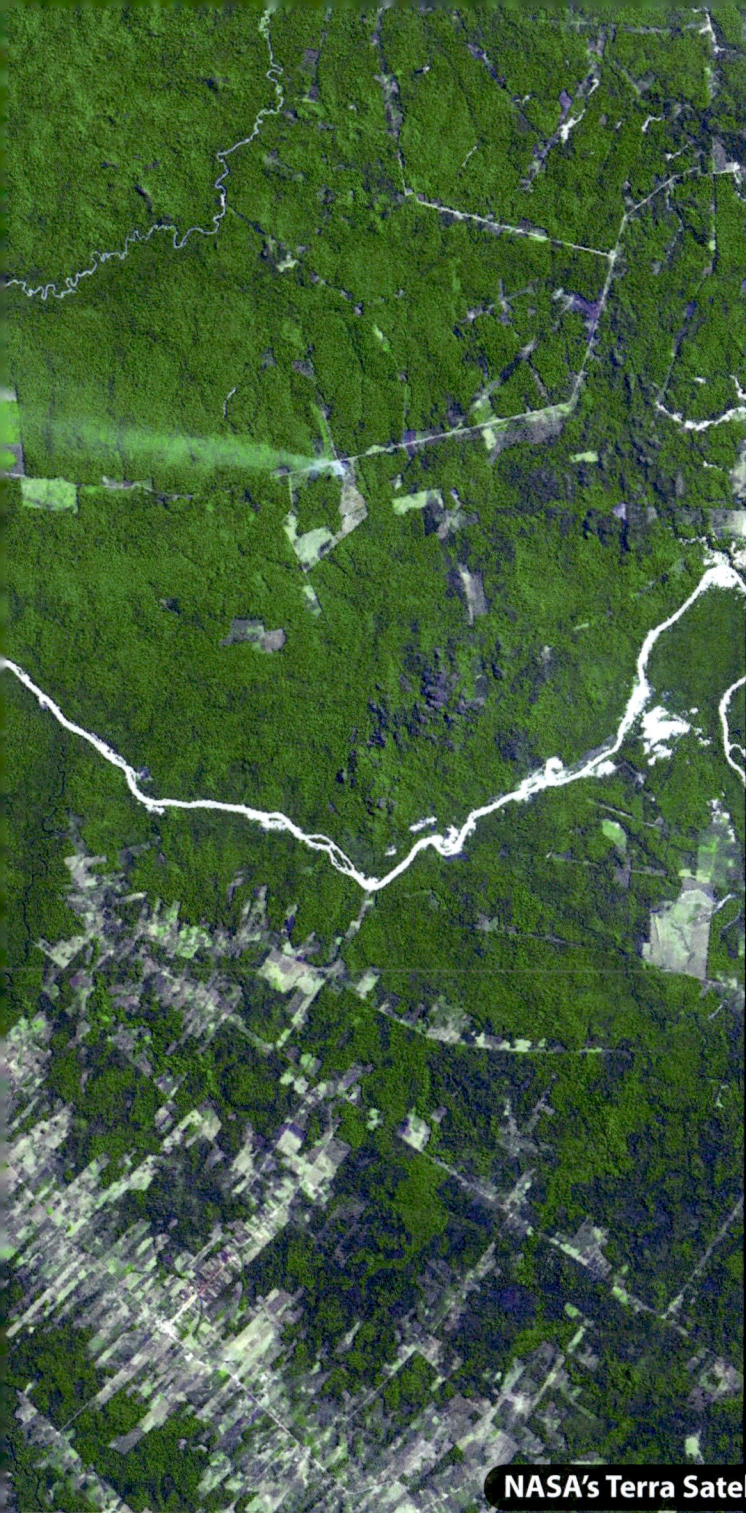

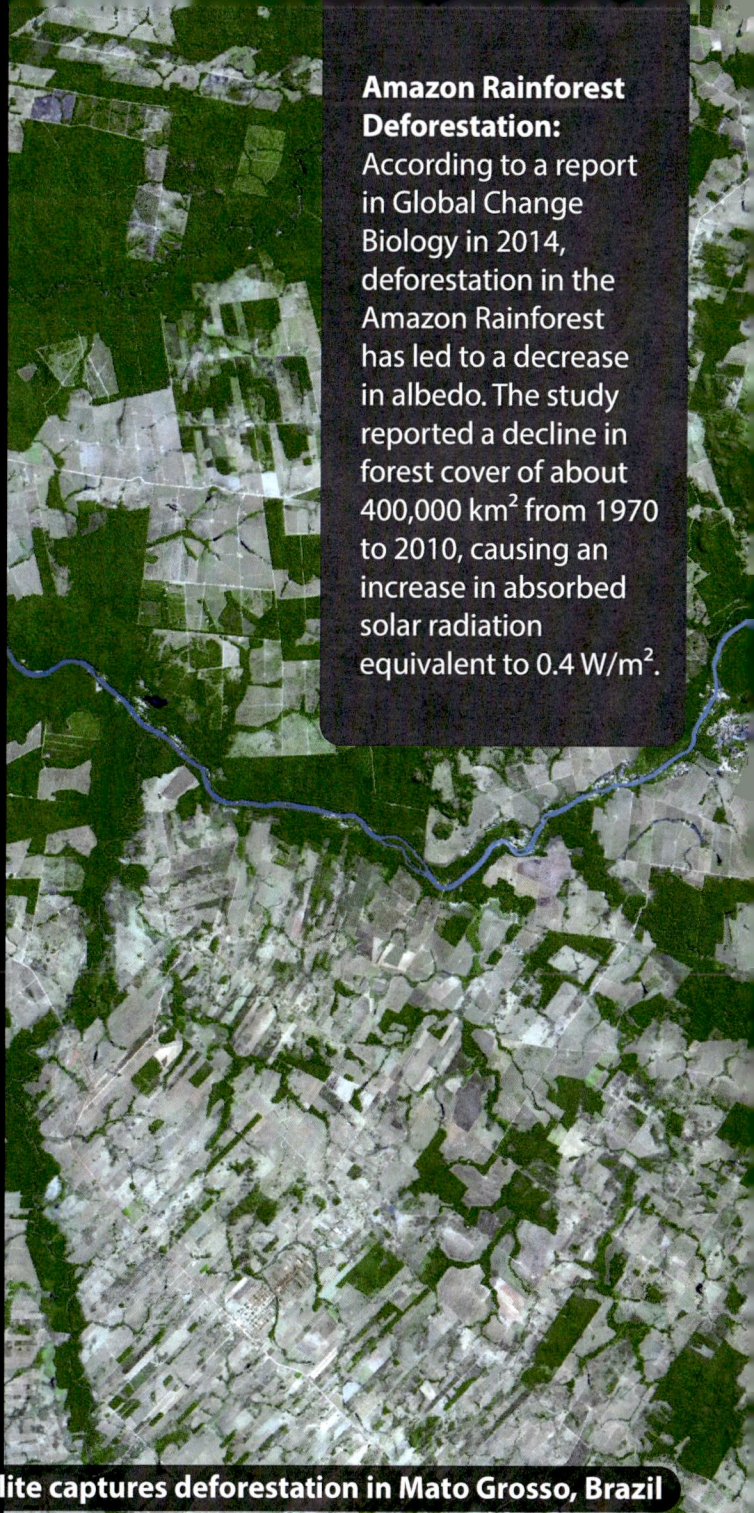

Amazon Rainforest Deforestation: According to a report in Global Change Biology in 2014, deforestation in the Amazon Rainforest has led to a decrease in albedo. The study reported a decline in forest cover of about 400,000 km² from 1970 to 2010, causing an increase in absorbed solar radiation equivalent to 0.4 W/m².

NASA's Terra Satellite captures deforestation in Mato Grosso, Brazil

Urban Heat Island Effect: The World Meteorological Organization noted that cities, due to the extensive use of dark materials like asphalt, typically have lower albedo than rural areas. A study in Los Angeles found a decrease in albedo of 0.20 (on a scale of 0 to 1) when transitioning from rural to urbanized areas. This change intensifies the urban heat island effect, raising city temperatures.

Jind
Population: 229,000

Rohtak
Population: 499,000

Bhiwani
Population: 268,000

NASA's Ecosystem Spaceborne Thermal Radiometer Experiment on Space Station instrument (ECOSTRESS)

Sonipat
Population: 382,000

Dehli
Population: 32,941,000

These examples underscore the importance of monitoring changes in albedo to manage climate change more effectively.

Each of these instances is also a testament to how human activities significantly influence Earth's albedo. By observing Earthlight, we can track these changes in albedo. This data enables us to monitor the advancement of climate change and provides crucial insights into the health of our planet.

Our climate models are reliant on precise albedo information. The fluctuations in Earth's albedo determine how much sunlight is absorbed by our planet compared to how much is reflected back into space. By keeping an eye on Earthlight, we can gain insights into these fluctuations, enabling us to refine our climate models.

Furthermore, Earthlight serves as a barometer for climate change. Variations in Earthlight's intensity, quality, or spread signal shifts in Earth's albedo and, consequently, our climate system. A dimming Earthlight, for instance, might signify melting polar ice caps, a prominent sign of climate change.

Nova Scotia, Prince Edward Island, and New Brunswick from the ISS

Recognizing this link between Earthlight and climate change is vital in our fight against global warming. It emphasizes the importance of maintaining our planet's delicate balance and equips us with the knowledge to protect it.

Sian's reflections

There isn't anything more fundamentally challenging for us as a species than understanding our changing climate and the role we play in the survival of so many species, including our own. It is part of the reason why I became a geoscientist.

I've always wanted to understand how the world around me works. Why do we have clouds and weather patterns and so many different climates? As soon as sunlight hits the atmosphere it is transformed by the Earth and all the molecules that make up the air we breathe.

As humans terraform our planet, we cause a ripple effect with how that sunlight is transformed into EarthLight.

Maybe if humanity knew about EarthLight then it could cause a ripple effect similar to the Overview Effect felt by astronauts and we could finally become environmentally conscious stewards of our world.

Dr. Sian Proctor illuminated by Earthlight

LEO PROCTOR

Chapter 5

SEEING EARTH, SEEING OURSELVES

The Astronaut's Perspective

Seeing Earth is Seeing Ourselves

Alone in the vast expanse of space,
An astronaut in orbit drifts in place,

Gazing out at the starry night,
Awaiting the Earth's majestic first light.

Then, slowly but surely, it comes into view,
A brilliant bright ball, shining blue,

With swirls of white and green and brown,
A masterpiece of colors, like a painted crown.

The astronaut watches, breathless and still,
As the mist of her eyes, begin to fill.

She feels all the awe and wonder,
As she beholds her world, in all its splendor.

She sees the continents and oceans wide,
The polar caps and deserts, side by side,

With inquisitiveness at the way they all snuggle fit and trim,
A puzzle piece of beauty, shining as bright as a gem.

Lost in musing as she floats
She feels her soul scribbling notes

With a sense of joy at the purity of sight,
She devours every second of Earth's captivating light.

For though she's far from home, it's true,
She's never felt so close, so near, so new,

To the planet that gave life birth,
And all the paradoxes that make up the Earth.

So she lingers there, for just a while,
Beneath the gentle comfort of Earth's brilliant smile,

And feels a sense of peace and grace,
As she drifts away, into outer space.

Dr. Sian Proctor

The Overview Effect, a term initially proposed by space philosopher Frank White in 1987, represents a transformative cognitive shift observed in astronauts during spaceflight, typically while beholding the Earth from orbit. This profound revelation transcends ordinary understanding, culminating in an overwhelming realization that our Earth is but a delicate, vibrant sphere of life suspended in the immense void of space.

Astronauts invariably characterize this experience as a pivotal moment in their lives. Observing our home planet, devoid of borders or divisions, and realizing the interconnected web of life engenders a heightened sense of responsibility for safeguarding our common domicile. Astronauts often return from their celestial journeys armed with a renewed commitment to environmental conservation and a more profound comprehension of humanity's universal kinship.

The Overview Effect instigates profound paradigm shifts in astronauts' worldviews.

Several astronauts have ventured into political arenas or established philanthropic organizations to champion global unity and environmental preservation, spurred by their profound space-bound epiphanies.

These profound transformations underscore the acknowledgment of Earth as an interconnected system, humanity as a unified family, and the pressing imperative of preserving our common habitat.

Let's hear what astronauts have said about their experience seeing the Earth form space:

Tracy Caldwell Dyson in the Cupola of the International Space Station

"Orbiting Earth in the spaceship, I saw how beautiful our planet is. People, let us preserve and increase this beauty, not destroy it."

As the first human in space, Russian cosmonaut **Yuri Gagarin** had a unique perspective. During his historic flight in 1961, he spoke of Earth's breathtaking beauty, saying, "The Earth is blue...how wonderful. It is amazing." His journey had a significant cultural impact and sparked a global interest in space exploration. The profound effect of seeing Earth from space undoubtedly influenced his worldview and that of those who followed his journey.

"Americans, Asians, everyone who has seen it says the same thing, how unbelievably beautiful the Earth is and how very important it is to look after it. Our planet suffers from human activity, from fires, from war; we have to preserve it."

Valentina Tereshkova, the first woman in space and a Russian cosmonaut, has described her experiences in orbit as life-changing. During her mission, she orbited Earth 48 times and was deeply moved by the planet's beauty. Tereshkova noted the unity and interconnectedness that become apparent when viewing Earth from space, expressing the need for peace and environmental protection.

B612 Foundation: Taking responsability for our future, together.

Apollo 9 astronaut **Rusty Schweickart** described the Overview Effect as an intense realization of the interconnectedness of all life on Earth. During a spacewalk, Schweickart experienced a profound shift in his perspective, which he described as an "empathic" feeling—a sense of unity and shared destiny with all living beings on the planet. He returned from his mission with a strong belief in the need for planetary defense against threats such as asteroids, leading him to co-found the B612 Foundation, dedicated to defending Earth from asteroid impacts.

"When you go around [the Earth] in an hour and a half, you begin to recognize that your identity is with that whole thing! And that makes a change. And you look down there, and you can't imagine how many borders and boundaries you cross, again and again and again. And you don't even see them."

Rusty Schweickart on an EVA on Apollo 9

Earth is a "grand oasis in the vastness of space."

Apollo 8 and Apollo 13 astronaut **Jim Lovell** spoke about the profound impact of seeing Earth from lunar orbit. He described the Earth as a "grand oasis in the vastness of space," which deeply struck him. Lovell emphasized how this perspective made the world's problems seem trivial compared to the larger universe and further emphasized our shared responsibility in preserving our home planet.

Andes Mountains. Image captured during Gemini Titan 7 mission by Astronauts James A. Lovell Jr., pilot, and Frank Borman, command pilot

"The thing that really surprised me was that [Earth] projected an air of fragility. And why, I don't know. I don't know to this day. I had a feeling it's tiny, it's shiny, it's beautiful, it's home, and it's fragile."

Lunar Modual Ascent Stage leaves the Moon

Apollo 11 astronaut **Michael Collins** has often spoken about his experience of the Overview Effect. Unlike Neil Armstrong and Buzz Aldrin, Collins never walked on the moon; he orbited it alone in the Command Module. During this time, he reflected on Earth's delicate beauty and vulnerability, stating, "The thing that really surprised me was that [Earth] projected an air of fragility. And why, I don't know. I don't know to this day. I had a feeling it's tiny, it's shiny, it's beautiful, it's home, and it's fragile."

An "explosion of awareness" and a "sudden recognition of the universe."

"Suddenly, from behind the rim of the moon, in long, slow-motion moments of immense majesty, there emerges a sparkling blue and white jewel, a light, delicate sky-blue sphere laced with slowly swirling veils of white, rising gradually like a small pearl in a thick sea of black mystery. It takes more than a moment to fully realize this is Earth... home."

Edgar Mitchell, Apollo 14 astronaut, famously described his experience with the Overview Effect. Mitchell's journey back to Earth following the mission culminated in a profound epiphany, where he experienced an intense sense of universal connectedness—an epiphany that led him to establish the Institute of Noetic Sciences. He described his experience as an "explosion of awareness" and a "sudden recognition of the universe." Mitchell came to realize that the molecules in his body, the spacecraft, and his companions had been manufactured in the furnace of ancient stars.

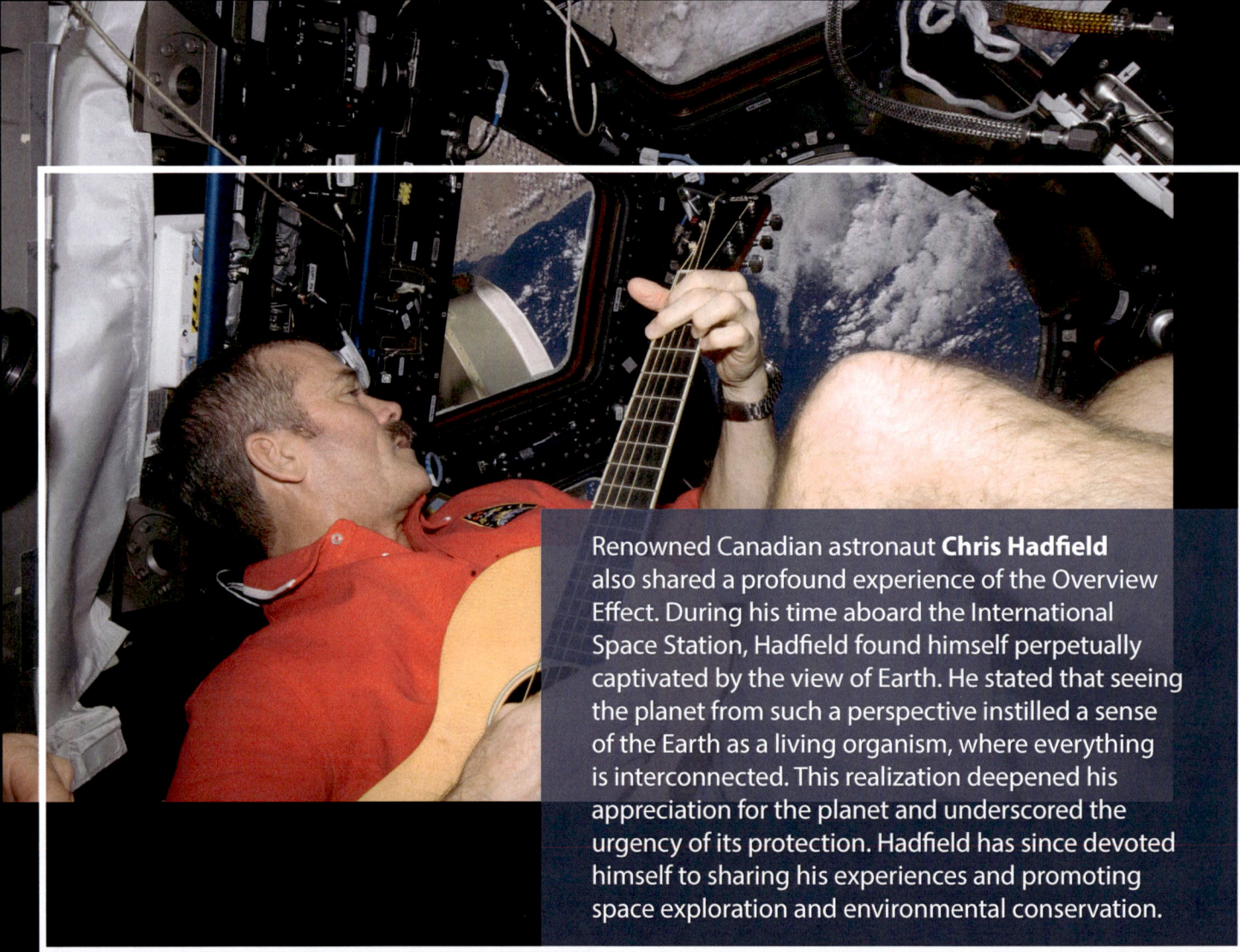

Renowned Canadian astronaut **Chris Hadfield** also shared a profound experience of the Overview Effect. During his time aboard the International Space Station, Hadfield found himself perpetually captivated by the view of Earth. He stated that seeing the planet from such a perspective instilled a sense of the Earth as a living organism, where everything is interconnected. This realization deepened his appreciation for the planet and underscored the urgency of its protection. Hadfield has since devoted himself to sharing his experiences and promoting space exploration and environmental conservation.

"Most people see the world in the area they're familiar with, and then the rest of it is sort of theoretical, or a distant unknown. To go around the world thousands of times, like I have, you really get to know the world for what it truly is, with its infinite variety and its amazing ability to self-heal, but also the damage that we're inflicting to it."

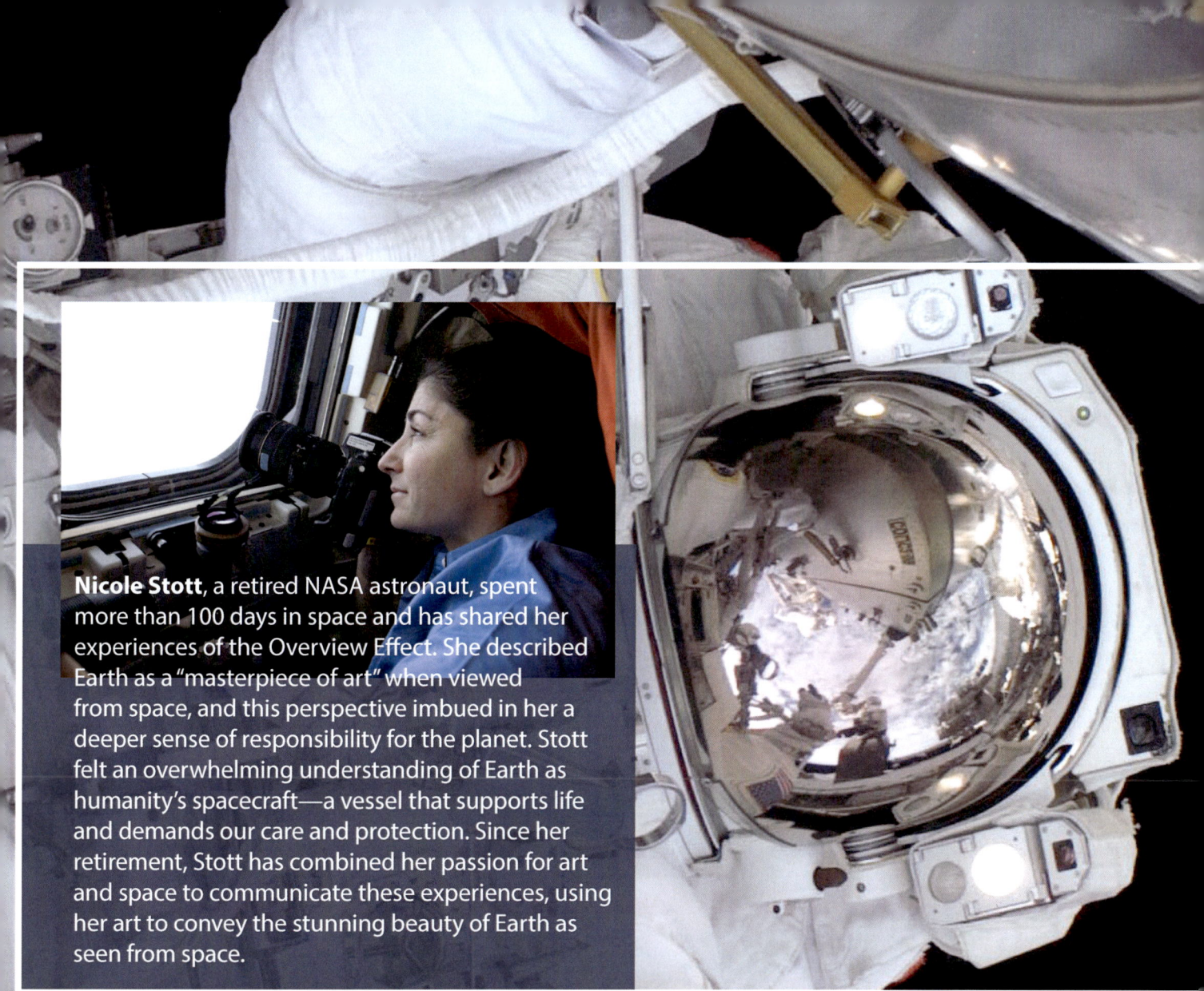

Nicole Stott, a retired NASA astronaut, spent more than 100 days in space and has shared her experiences of the Overview Effect. She described Earth as a "masterpiece of art" when viewed from space, and this perspective imbued in her a deeper sense of responsibility for the planet. Stott felt an overwhelming understanding of Earth as humanity's spacecraft—a vessel that supports life and demands our care and protection. Since her retirement, Stott has combined her passion for art and space to communicate these experiences, using her art to convey the stunning beauty of Earth as seen from space.

"We live on a planet! The best planet. This truth became clear to me as I circled the Earth on the space station, and from that special vantage point I was presented with the reality of Earth as our home."

SPACE FOR ART
FOUNDATION

"There's no greater beauty than looking at the Earth from up high – and I'll never forget the first time I saw it. After take-off we left the atmosphere and suddenly light streamed in through the window. We were over the Pacific Ocean. The gloriously deep blue seas took my breath away."

Helen Sharman, the first British astronaut and the first woman to visit the Mir space station, has reflected on her experiences with the Overview Effect. She noted that observing Earth from space emphasizes its fragility, prompting thoughts about the importance of conservation. Sharman also highlighted the absence of political boundaries, reinforcing a sense of global unity and shared responsibility for our planet.

"You see the arc of the Earth and you see the thin layer of atmosphere, this shiny, shimmering blue. The images that are most vivid for me are going across the Horn of Africa, crossing the Nile Delta. And there's this just shimmering light that comes off of the desert."

Dr. Mae Jemison, the first African-American woman in space, has expressed how her spaceflight experience on the Space Shuttle Endeavour influenced her perspective of Earth. She spoke of Earth as a planet, part of a system, not a solitary entity. Viewing Earth against the backdrop of the universe, she felt both a sense of peace and a realization of the fragility of our planetary ecosystem. She was struck by the thinness of the atmosphere, and upon her return, she became an advocate for the careful stewardship of our shared environment.

"As a species, we're so temporary, transient — we could be gone and the Earth would just keep on moving. There's nothing permanent or inevitable about us."

Samantha Cristoforetti, first Italian woman in space

"We are citizens of space, and stewards of Earth. We need to take actions to build [a] global climate alliance in order to protect our environment."

Soichi Noguchi, Japanese astronaut

> **"[The best part of going to space is] seeing Earth. It's breathtaking and beautiful."**
>
> **Charles Bolden Jr.**

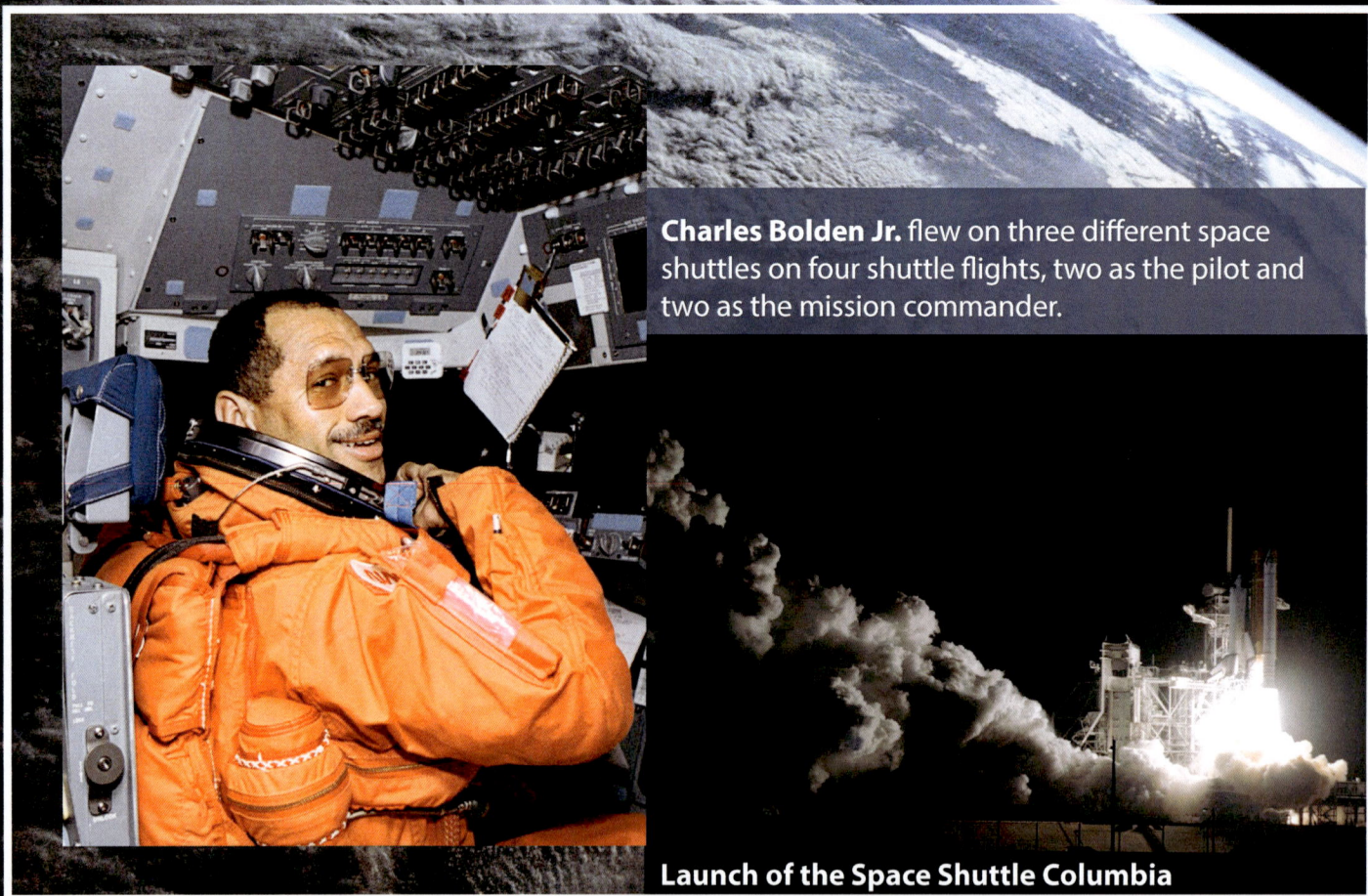

Charles Bolden Jr. flew on three different space shuttles on four shuttle flights, two as the pilot and two as the mission commander.

Launch of the Space Shuttle Columbia

Photo taken from Space Shuttle Discovery over the Andes Mountains and the Pacific Ocean

"The Earth was small, light blue, and so touchingly alone. Our home that must be defended like a holy relic. The Earth was absolutely round. I believe I never knew what the word 'round' meant until I saw Earth from space."

Alexei Leonov, first person to conduct a spacewalk

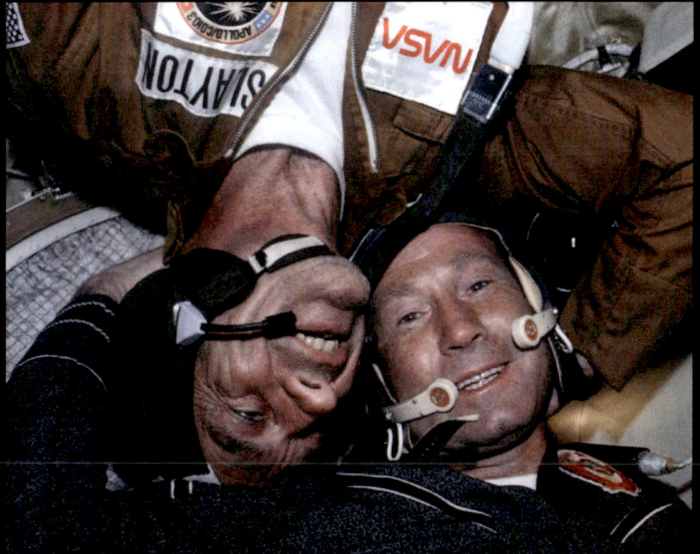

Astronaut Donald Slayton and Cosmonaut Alexei Leonov after a successful docking of the Soyuz and Apollo spacecrafts.

Soviet Soyuz 19 while on approach to dock with an Apollo spacecraft as part of the Apollo-Soyuz Test Project.

"Every single part of the Earth reacts with every other part. It's one thing. Every little animal is important in that ecosystem. [Seeing the Earth] makes you realize that, and makes you want to be a little more proactive in keeping it that way. If I could get every Earthling to do one circle of the Earth, I think things would run a little differently."

Karen Nyberg

"I'll never forget the views of Earth, how you could see the curvature of the Earth, the sun rising and setting very quickly. You could see all the different kinds of geographical features on the ground, deserts, mountains, forests and oceans. It was really stunning, just breathtaking. I could have done nothing for the two weeks except look outside."

Nicholas Patrick

Nicholas Patrick on EVA

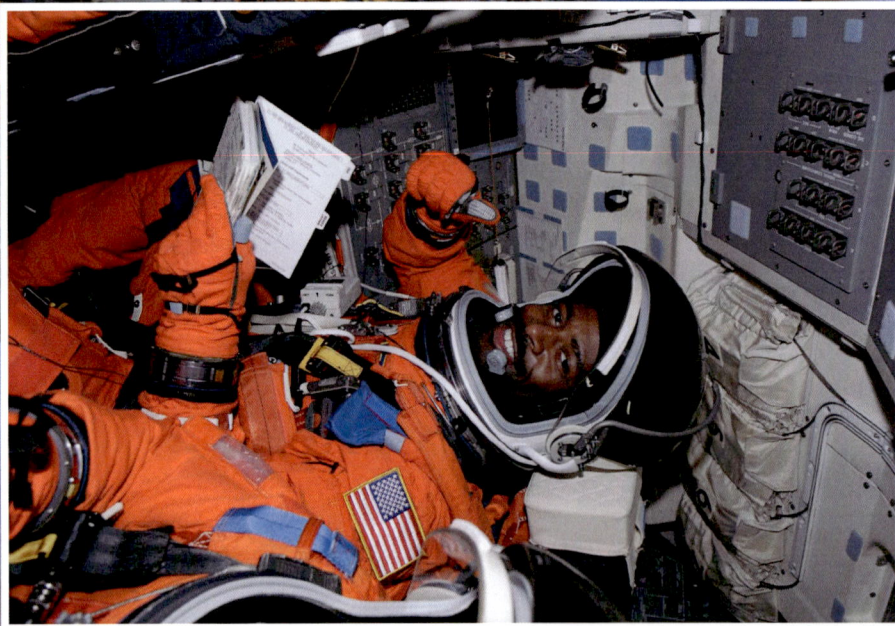

"I had that cognitive shift, that moment of, 'Oh my god, we're all connected. Oh my god, the planet is so beautiful. Oh my god, what can I do to make it better when I get home?' And in coming home, the things that bothered you? They don't bother you."

Leland D. Melvin

"We just watched the world go by together, often in stunned silence, amazed at how beautiful our homeland is."

NASA astronaut Jessica Watkins over the Atlantic Ocean and South Africa

"I hope I never recover from this."

"I knew that what I was feeling was something I will carry with me for the rest of my life, something I must share with as many people as I can. Going into space made me so aware of how fragile our lives are here on Earth, how we *need* each other, and need to continue to strengthen the bonds that connect us to each other. Because out there, there is no life. There is no *us*."

William Shatner

"Seeing the planet for the first time is like being born once again. You don't know how to react, what to say, what to think what to feel. All you know is life is not what you thought it was and your home is calling you back."

Katya Echazarreta

"Once I finished operating my science payload, I finally turned my full attention to planet Earth — seeing it with my own eyes, and truly experiencing it as a planet — was profound. I was looking down at everything I've ever known and everyone I've ever loved and I was completely struck by the fragility of it all."

Kellie Gerardi

"Getting a glimpse of the Earth from that perspective was one of the most awe-inspiring experiences of my life! Looking at it, it becomes so clear how small we are in every sense: while we're down here on solid ground, fighting amongst ourselves over petty matters, there's still a whole literal universe waiting to be explored!"

Victor Hespanha

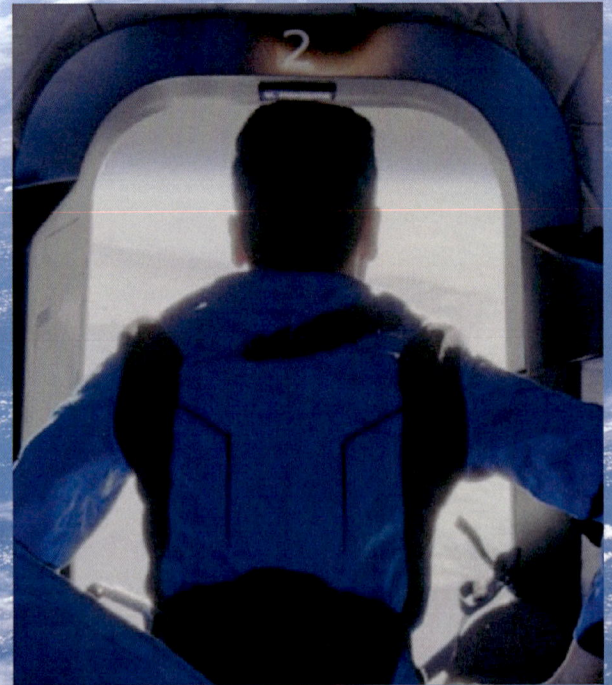

"We have all seen pictures of Earth taken from space, but nothing quite prepares you for that first encounter through the window. Inspiration4 orbited at an altitude where I physically felt a very real divide between us and the atmosphere with that black vacuum of space creating an obvious separation. However, watching the planet go by, seeing shadows lengthen, observing snowy mountaintops next to perfectly blue ocean waters, experiencing thunderstorms put on a spectacular lightning display—I have never felt more connected to our home."

Chris Sembroski

The Overview Effect Transforming Humanity's Vision

The opportunity to experience the Overview Effect, a deep transformation of worldview often reported by astronauts, is on the verge of becoming available to many. This change involves seeing Earth not as a collection of individual countries but as a fragile, interconnected whole. As more people gain access to space travel, the Overview Effect could revolutionize how we perceive our planet and ourselves.

The commercialization of space travel through companies such as SpaceX, Blue Origin, and Virgin Galactic has opened the gates of space to the public. This 'democratization' of space could allow the profound impact of the Overview Effect to permeate throughout society.

Imagine a world where this shift in perspective becomes a common experience. We could see a surge in efforts towards preserving our environment, increased collaboration across nations, and a renewed dedication to peace. This shared experience could stimulate a global interest in science and technology, inspiring future generations to keep exploring the cosmos.

Sian's reflections

I went to space fully knowing about Frank White and the Overview Effect and wondered if I would have some profound transformative shift while staring back at Earth from space. The result is YES! I was moved and transformed by both the view and EarthLight. It was stunningly bright and transfixing. I was immediately hooked by EarthLight and, like Leonardo, wanted to know more.

Could this be a key component to why astronauts experience the Overview Effect in space but when people on earth are shown the earth from that same perspective aren't transformed?

When I came home and tried to explain the experience to family and friends on earth, all I could say to get them to even slightly understand what I

Ultimately, the Overview Effect could be the spark that ignites a global shift in consciousness. Seeing the vulnerability and beauty of our shared home from a vantage point could lead us to rise above our differences and unite for a sustainable future.

had experienced was to relate it to moonlight. I'd say, you know when you go outside at night and you see a full Moon rising, and you can walk in moonlight.

Think about how that makes you feel. Think about how humanity has been mesmerized by the moon since we began to walk upright. Think about all the myth, lore, love, and even werewolf's that have been woven into the culture and psychology of humanity. Earthlight is like that but only more intense and spectacular.

That's when the light bulb goes off and people begin to understand the beauty of EarthLight.

Image: SpaceX Crew 2 mission

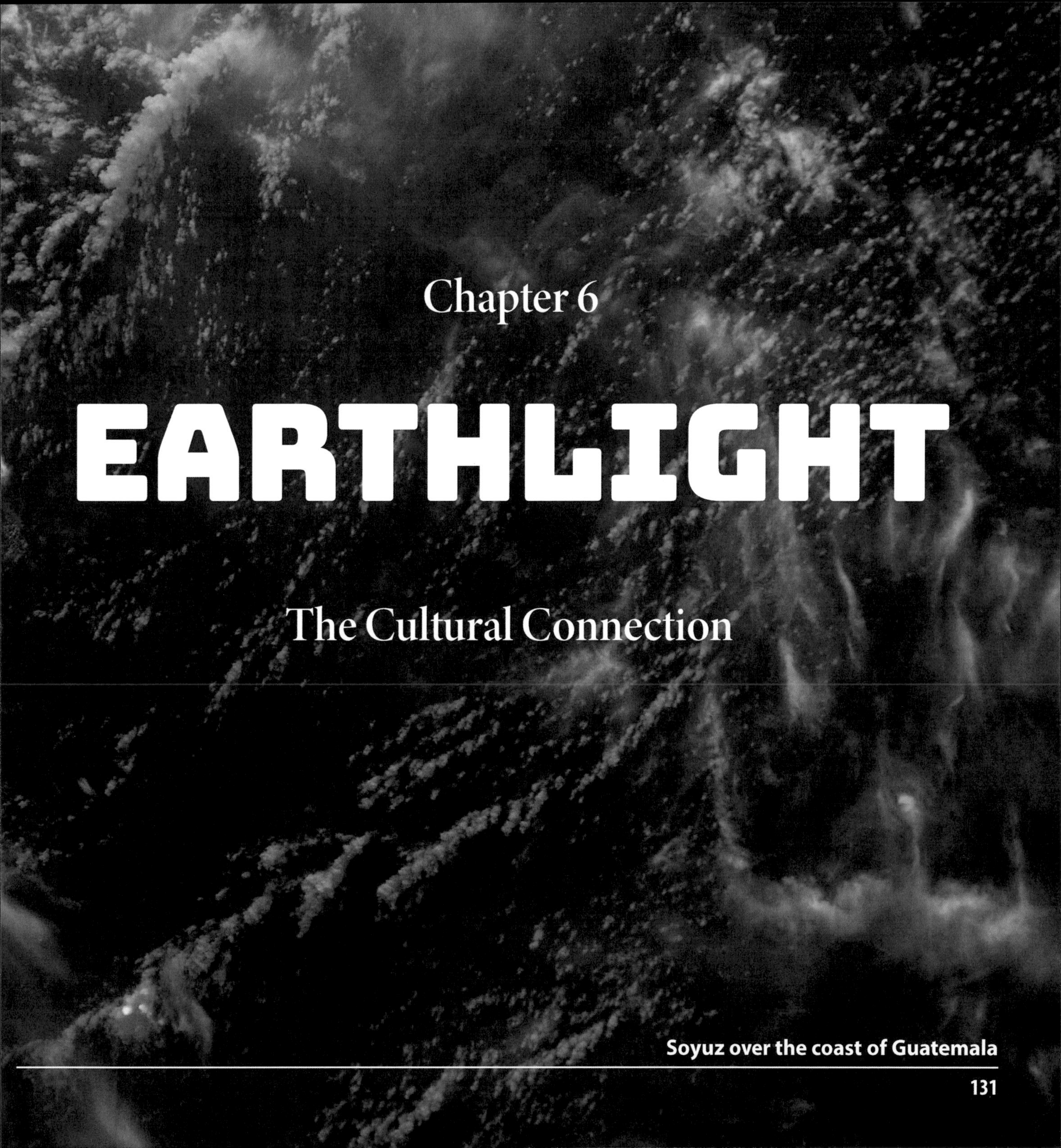

Chapter 6

EARTHLIGHT

The Cultural Connection

Soyuz over the coast of Guatemala

EarthLight – the cultural connection

Do you belong to me
This enchanted relationship of reckless love
Always wanting more
Yet not willing to give way

I long to feel your embrace
While shadowed by self-preservation
Who will remember us
If not you

Begging to be beloved
Seeing first hand your unrelenting beauty
No recourse but to give way
Bow down in forgiveness

The spear tip poisoned
By the child at play
Let thy love last forever
In the heart of our existence

I surrender

Dr. Sian Proctor

Earthlight holds a prominent place as a muse in the art of photography. Its delicate spectrum of colors adds intricacy and detail to photographs, enhancing their visual allure. Photographs of our planet from space are mesmerizing. Earthlight adds a new dimension to the landscapes, casting dramatic shadows, and emphasizing Earth's geographical features. The resulting images hold not just scientific value, but also offer an irresistible draw for the art-loving eye. Here are examples of how Earthlight has been captured on Earth and beyond.

Above the Sea of Okhotsk in the western Pacific Ocean

Noctilucent Clouds: This image was taken from the ISS during expedition 67. Noctilucent clouds are ice crystals that form at around 85 kilometers above Earth's surface.

We came all this way to explore the moon and the most important thing is that we discovered the Earth.

Bill Anders

Earthrise

Perhaps the most famous example of Earthlight captured in a photograph is the "Earthrise" image taken by astronaut William Anders during the Apollo 8 mission in 1968. The photo shows a stunning view of Earth emerging from the lunar horizon, bathed in Earthlight.

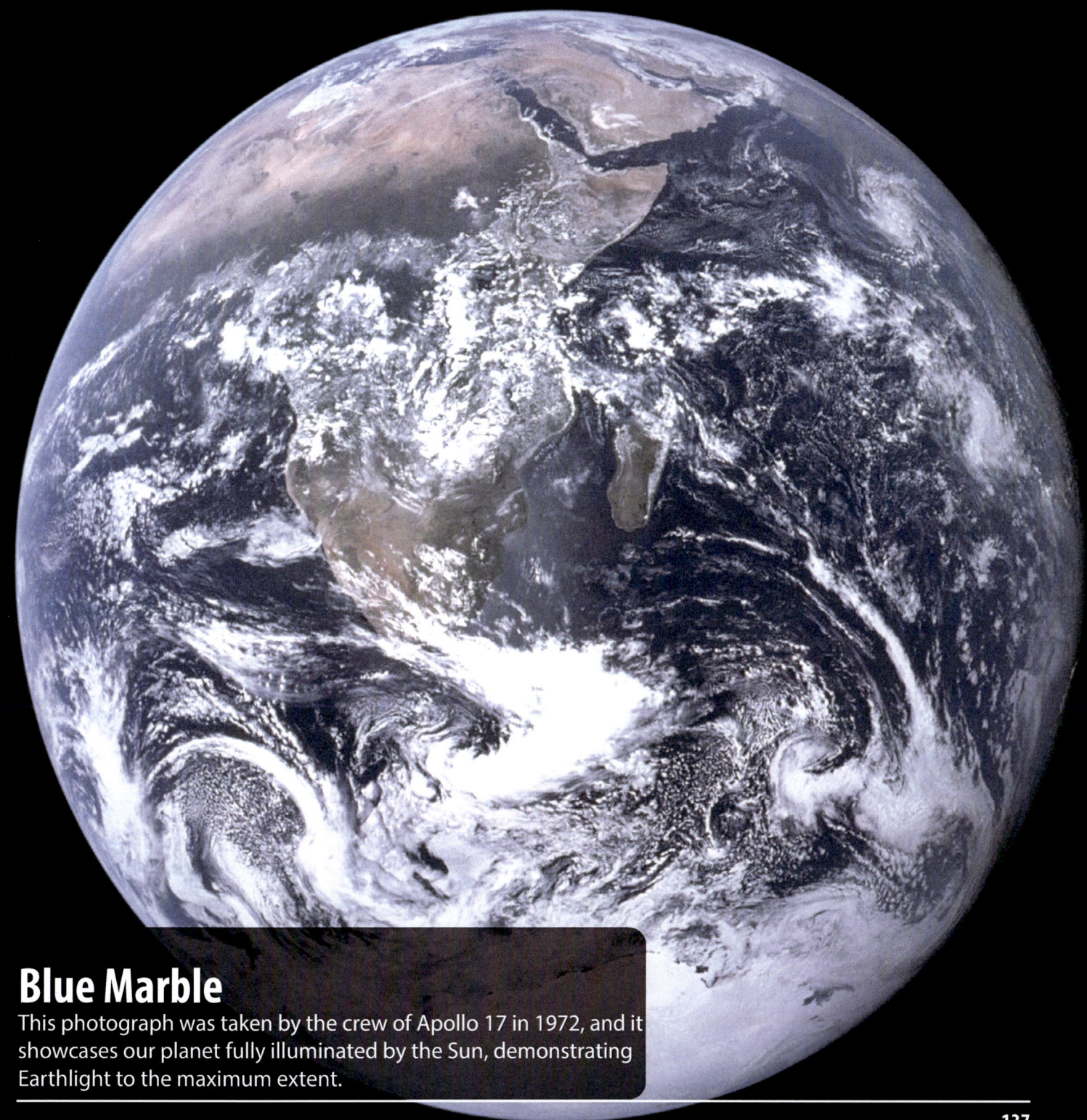

Blue Marble

This photograph was taken by the crew of Apollo 17 in 1972, and it showcases our planet fully illuminated by the Sun, demonstrating Earthlight to the maximum extent.

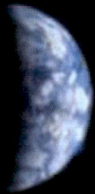

Crescent Earth

An image captured by the Mars Orbiter Camera (MOC) on NASA's Mars Global Surveyor, shows Earth as a crescent, much like the Moon viewed from Earth, showcasing Earthlight on the darkened portion of our planet.

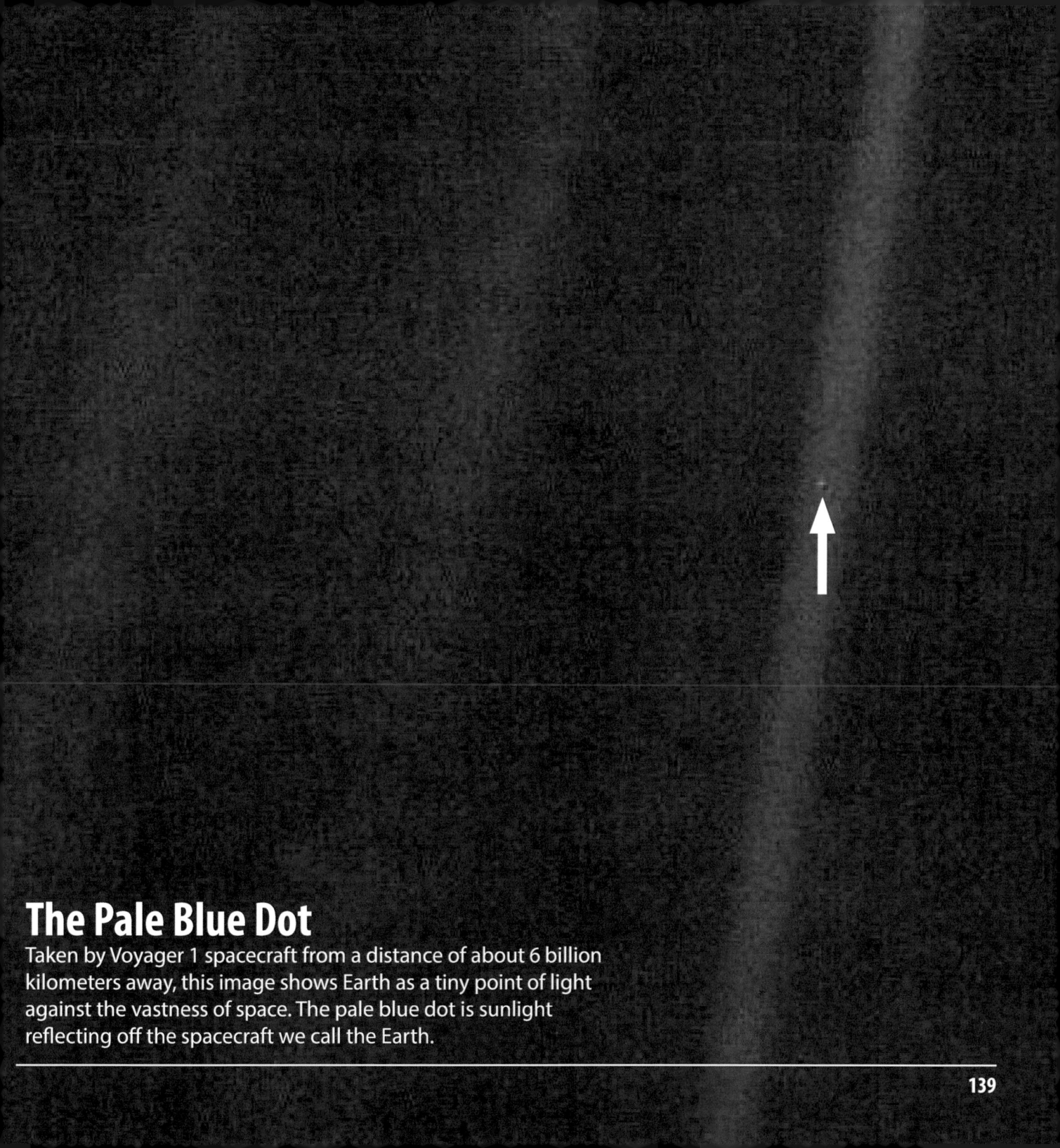

The Pale Blue Dot

Taken by Voyager 1 spacecraft from a distance of about 6 billion kilometers away, this image shows Earth as a tiny point of light against the vastness of space. The pale blue dot is sunlight reflecting off the spacecraft we call the Earth.

Earthlight on the International Space Station (ISS)

Astronauts aboard the ISS often capture stunning photos of Earth. Earthlight can be seen illuminating the underbelly of the ISS.

Earthlight on Other Celestial Bodies

Earthlight illuminates the dark side of the Moon, a phenomenon known as earthshine. Images of the Moon with the dark side subtly visible due to earthshine provide real-world examples of Earthlight.

Crescent Venus

Venus, when seen from Earth as a thin crescent, is also a subject of Earthlight. Though impreceptable from Earth, the nightime side of Venus is faintly illuminated by Earthlight. An astronaut living above Venus's clouds, could probably observe a faint shadow when Earth is overhead in her sky.

Earthlight in Star Trails
Long exposure photographs that capture star trails also often show the subtle glow of Earthlight from the atmosphere.

Total Lunar Eclipses

During a total lunar eclipse, the Moon takes on a red hue due to sunlight being refracted by Earth's atmosphere,.

Astrophotography

Nighttime landscapes with the Milky Way or other celestial bodies in the backdrop often also capture Earthlight, especially when there is a new moon. The faint glow in the sky and on the landscape is Earthlight.

These iconic images have had a profound impact on humans in many ways:

Promoting Global Unity and Environmental Awareness:

Images like "Earthrise" and "Blue Marble" provide a planetary perspective that underscores Earth's fragility and the need to protect it. Viewing Earth as a small, solitary planet in the vast expanse of space can generate a sense of unity among all people and a desire to take care of our shared home.

After an orange cloud — formed as a result of a dust storm over the Sahara and caught up by air currents — reached the Philippines and settled there with rain, I understood that we are all sailing in the same boat."

Vladimir Kovalyov

Dust storm in the Sahara

Highlighting Exploration and Progress:

These photos also symbolize human achievement. They are reminders of our capability to reach beyond our planet and explore the cosmos, which can inspire continued advancements in technology, science, and space exploration.

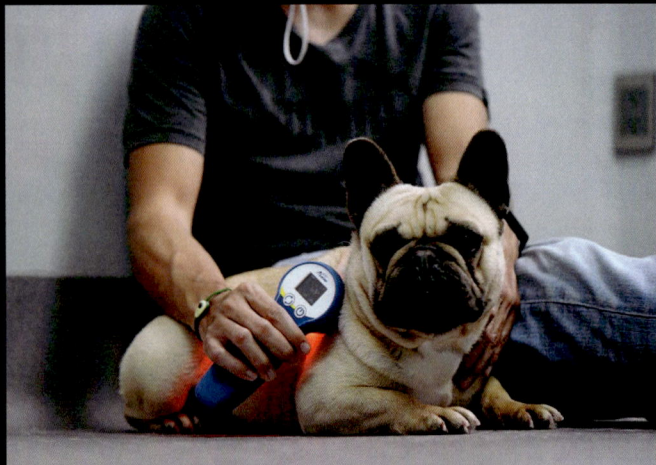

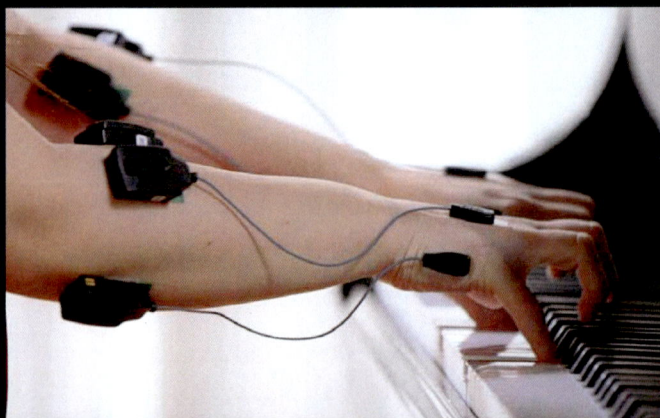

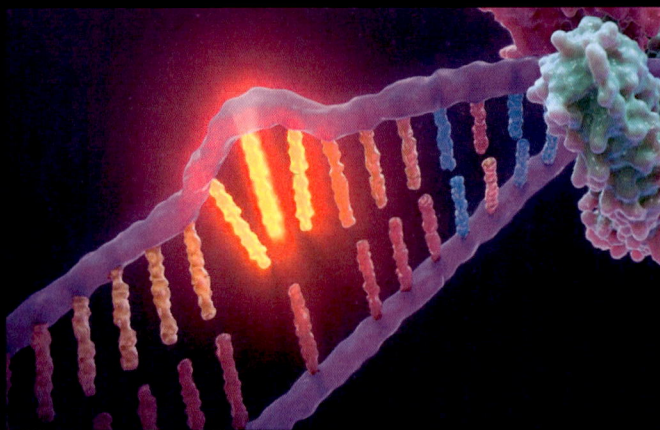

Only a few of the technologies from the NASA Spinoff 2023 magazine

Promoting Space Research and Funding:

Iconic Earth images from space have arguably spurred interest and investment in space exploration. They help the public visualize the importance and outcomes of space research, potentially encouraging greater funding and support.

Genes in Space-10 winner, High School Senior, Pristine Onuoha with her investigation on stem cells

"I realized up there that our planet is not infinite. It's fragile. That may not be obvious to a lot of folks, and it's tough that people are fighting each other here on Earth instead of trying to get together and live on this planet. We look pretty vulnerable in the darkness of space."

Alan Shepard

Provoking Philosophical Thought:

Images such as "The Pale Blue Dot" encourage deep philosophical contemplation. They remind us of our minuscule place within the cosmos, inspiring humility and a sense of wonder.

Coastal Brazil, Atlantic Ocean, West Africa, Sahara, Antarctica

Artist Concept of Astronaut on Mars

Advancing Scientific Understanding:

Images showing Earthlight advance our understanding of the solar system. They provide valuable data for scientists studying everything from lunar geology to climate patterns on Earth.

**Images taken by Middle Schoolers using the cameras on
Space Shuttle Atlantis of the Fires in Indonesia**

Education:

These images are widely used in educational settings, helping to illustrate concepts in earth science, astronomy, and other fields. They've undoubtedly played a role in sparking the curiosity of countless students, possibly inspiring the next generation of astronomers, astronauts, and earth scientists.

Students take part in NASA's JASON project examining the loss of wetlands in the Louisana bayou

NASA astronaut Leland Melvin and Sesame Street's Elmo talk about living on Orbit

NASA astronaut Randy Bresnik, left, and ESA astronaut Paolo Nespoli answer questions at the Washington School for Girls

In essence, these photographs have become cultural and scientific touchstones, inspiring awe, sparking curiosity, and reminding us of our shared responsibility to care for our planet.

Earthlight underscores the fact that photography transcends mere capture; it is about crafting an image that narrates a story, stirs emotions, and stimulates thought.

Earthlight
is the light you see around you everyday.

Earthlight
is expressed as the beautiful colors making up the world around us.

Earthlight
The Heartbeat of
Human Expression

Earthlight, the gentle luminescence mirrored back from our home planet, beats in rhythm with the heart of human expression. Its influence weaves itself through the canvas of art, the melody of music, the lyricism of poetry, and the expansive narratives of literature.

Earthlight is Art

Capturing the essence of Earthlight involves creating works that evoke the emotional and psychological impacts of the phenomena. Artists can depict scenes of Earth as viewed from space, highlighting the contrast of our vibrant planet against the darkness of space. They immortalize Earthlight in hues and shades. Paintings, photographs, and installations harness the allure of Earthlight, urging a deeper kinship with the cosmos and a contemplation of our position within it.

Earthlight is Poetry

In poetry, Earthlight and the Overview Effect can be captured through vivid, evocative language that draws readers into the experience of seeing Earth from space. Poets use metaphors and similes to relate these extraordinary experiences to more familiar ones. The rhythm and flow of the poetry mimics the peaceful, floating sensation of space travel and the gentle glow of Earthlight.

Poems explore themes of unity, interconnectedness, and environmental stewardship, reflecting the profound realizations associated with the Overview Effect. The silvery glow of Earthlight lends itself to verses rich in tranquility. The poetic imagery, a gift from the cosmic perspective, finds a home in lines that transport the reader into serenity, nurturing a profound celestial appreciation.

Music, the universal

Earthlight in Music

language, has also danced with Earthlight. Compositions have echoed its subtle charm, inspiring feelings of ethereal solitude and transcendence.

Through this celestial harmony, Earthlight bridges the human and cosmic experience, creating an emotive resonance that is universally understood. Composers create symphonic works and atmospheric ambient pieces that evoke the vastness of space and the fragile beauty of our planet. Just listen to astronomer Will Young's "Deep Sky Tunes" and you'll hear the inspiration.

Music can also incorporate sounds derived from space (like pulsar rhythms or solar wind sonifications) to heighten the celestial atmosphere.

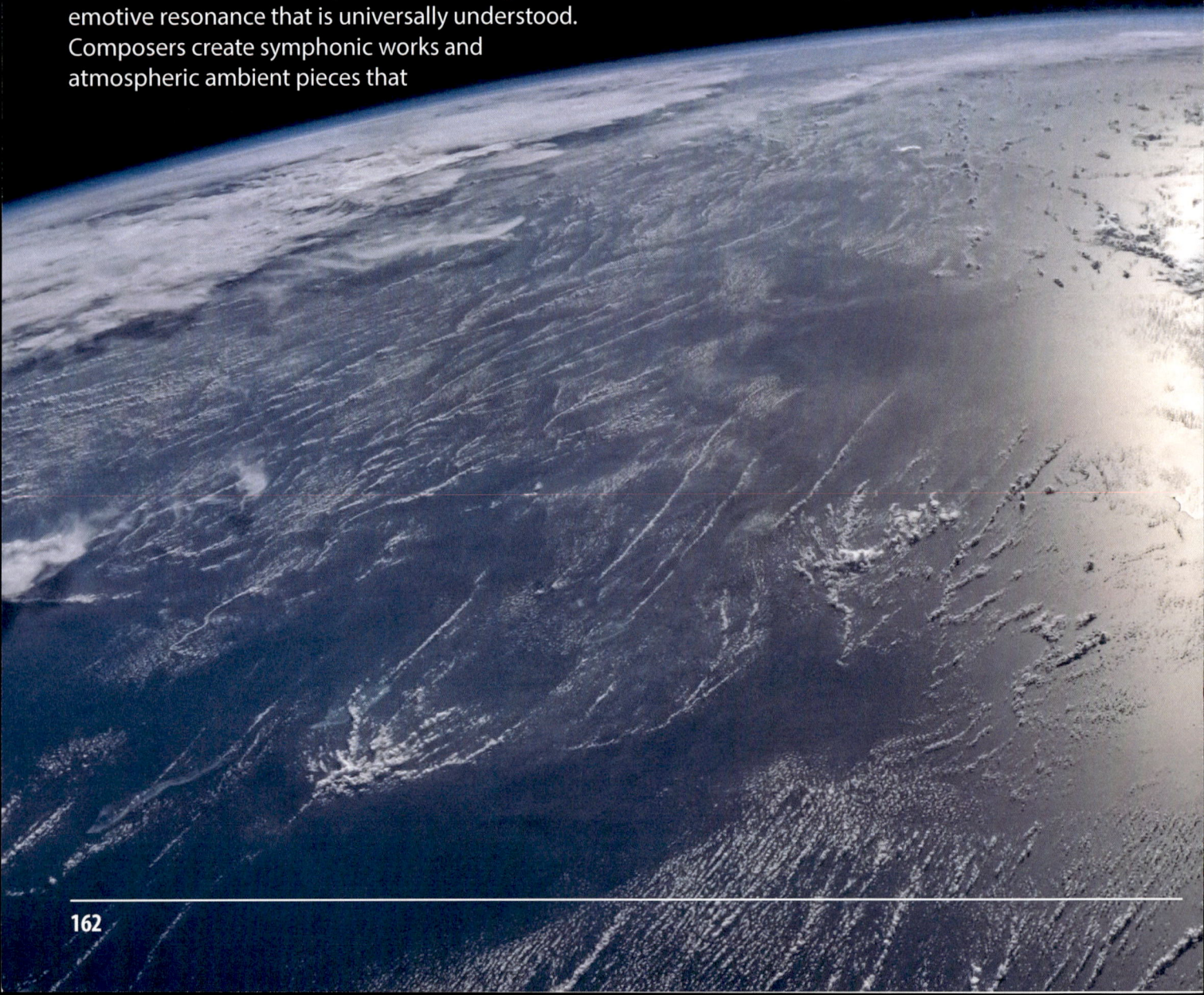

Just listen to Wyclef Jean's epic song, Borrowed Time, which uses the actual radio signature of Jupiter in its verses.

Earthlight in Film

In film, directors and cinematographers utilize visual effects and cinematography to depict Earthlight and the Overview Effect. This can involve scenes shot from a 'space' perspective, showing Earth from above, like in the movie Interstellar where the launch sequence is pared with music and later, absolute silence to heighten the effect.

These narratives explore astronauts' experiences and transformations, mirroring the psychological shifts associated with the Overview Effect. Documentaries utilize actual footage to convey these phenomena directly, in hopes that the viewer will feel something akin to the Overview Effect.

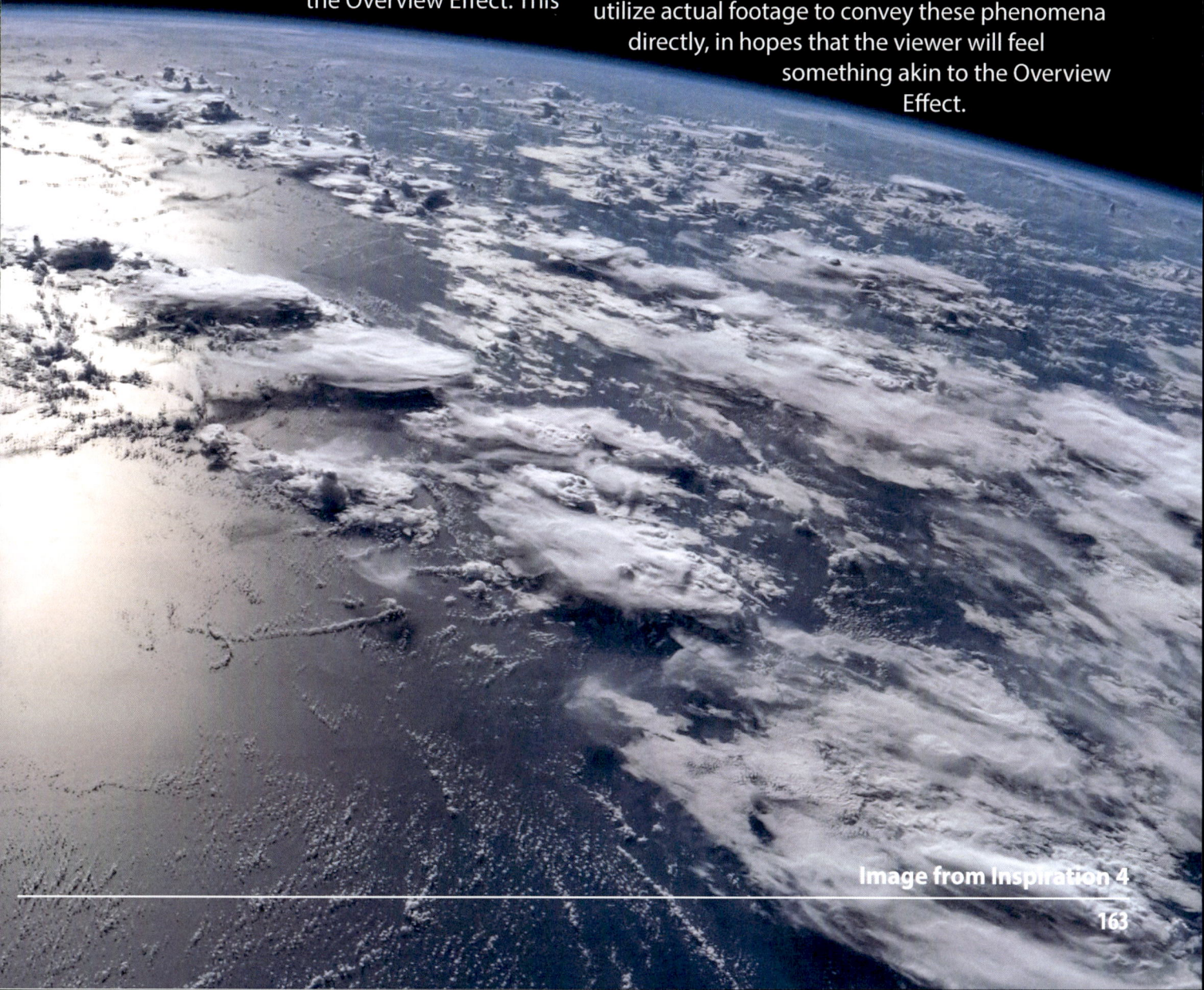

Image from Inspiration 4

Earthlight has inspired authors to create worlds...

Literature, too, has embraced Earthlight. Its presence within stories illuminates characters' journeys, signifying hope, enlightenment, and introspection. Science fiction, in particular, has drawn on the ethereal nature of Earthlight. Descriptive language can paint vivid pictures of Earthlight, evoking the awe-inspiring experience of seeing our planet from a distance. The Martian Chronicles" by Ray Bradbury: Bradbury's classic sci-fi work frequently references Earthlight as seen from Mars, impacting the atmosphere and the storyline.

(Image: Cover Grand Master Edition of 'The Martian Chronicles', by Michael Whelan)

Colors of Mars
Amateur space-photo editor Jason Major combined frames of different wavelengths from a Mars Rover's monocrome cameras to create this color image of the Martian landscape.

Marslight

The Red Planet, Mars, has etched a powerful impression on humanity, even though its reflected light, known as Marslight, has yet to touch a human face from less than tens of millions of kilometers.

Regardless, the planet's distinctive crimson in the sky has fired our imagination, influenced culture and incited scientific exploration.

Marslight has shaped our understanding of the planet indirectly, through images sent by Mars rovers that capture the planet's unique landscape. These glimpses feed our collective curiosity about the planet's geology, climate, and potential for life.

With crewed Mars missions on the horizon, we can only speculate about the experience of Marslight up close. Given Mars's distance from the Sun, the intensity of the reflected sunlight would be much fainter that Earth's.

Still, its psychological impact might be significant, echoing the transformative Overview Effect astronauts experience when they first see Earth from space. The experience of Marslight will mark a new phase in our cosmic journey, prompting us to reevaluate our place in the universe.

Planetlight and the future of humanity

Over the centuries, the enigmatic glow of Earthlight has held humankind in its spell, prompting a range of theories, myths, and interpretations. From ancient Greek philosophers, who conceived the idea of the Moon as a reflective body, to Renaissance polymaths, like Leonardo da Vinci, who comprehended the science of Earthshine, our understanding of Earthlight has been a journey of discovery. But for humanity, this is only the beginning.

As we discussed throughout this book, Earthlight's influence extends beyond science, touching the realms of culture and spirituality. Every culture since the dawn of time has woven light's various reflections into their rich cosmologies.

In the future, these cultural realms are bound to expand to include all the worlds visited by humans and their decendents. The mythologies of the future are sure to include "Planetlight" from places beyond our wildest imaginations.

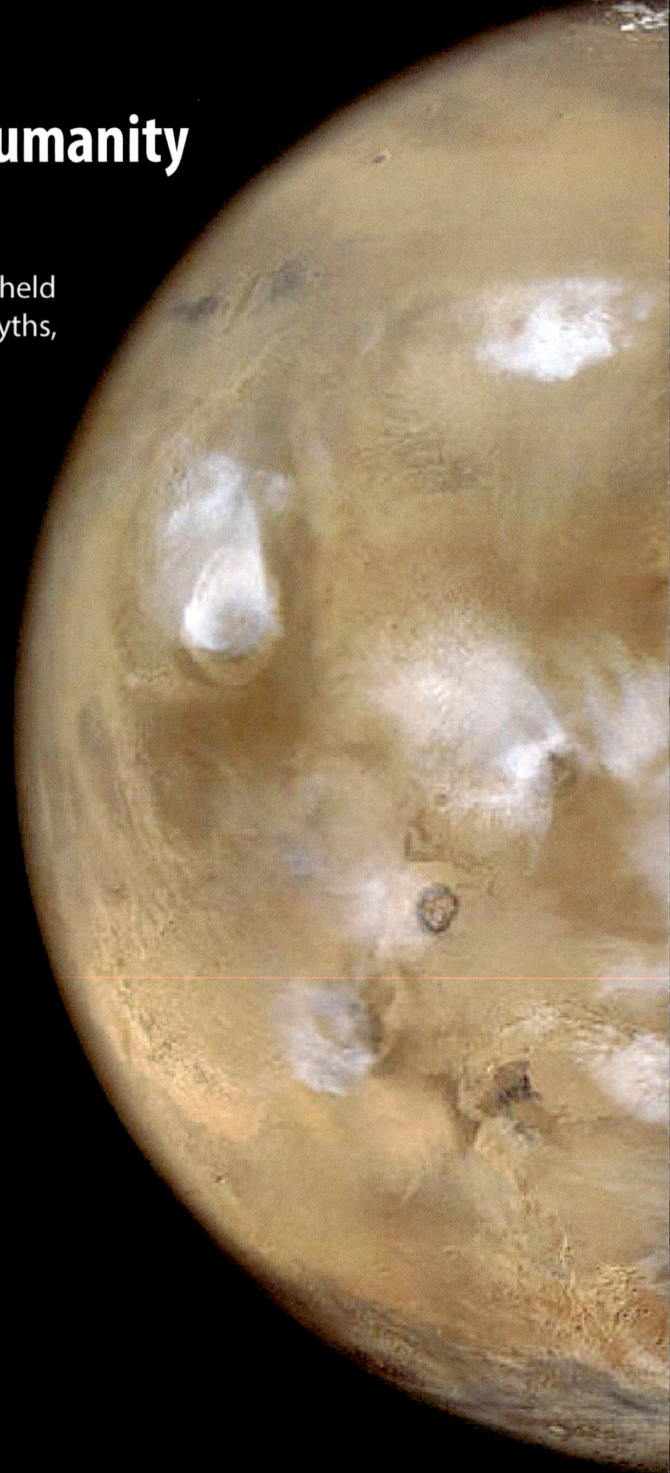

Sian's reflections

EarthLight is all around us. That's what I learned from going to space and is the message I bring home. Our planet is a giant prism that takes in sunlight and breaks it down into the beautiful world we see all around us.

We operate in a world where light is transformed and reflected. It's the way our planet communicates to us the properties that make up our world. We experience EarthLight every day when we walk outside and see the colors of the ground, the trees, the animals, the buildings, and the sky.

EarthLight stimulates our senses and stirs our imagination. It's why we see rainbows and sunsets. EarthLight is fundamental to our existence and culture in ways that are similar to sunlight and moonlight but because we are living in EarthLight we've lost the connection.

It is now time to get reconnected and open our eyes to the wonder and awe that is the EarthLight all around us. There is no black or white, there is only EarthLight and those are colors worth celebrating.

Glossary of Key Terms

Absorption: The process in which energy, such as light or heat, is taken in and retained by a substance.

Albedo: The measure of how well a surface reflects sunlight, often expressed as a percentage.

Angle of Incidence: The angle at which a ray of light or other waves strike a surface or interface.

Circadian Rhythm: The natural, internal biological clock that regulates an organism's sleep-wake cycle and other physiological processes over a 24-hour period.

Earthshine: The faint illumination of the Moon's surface caused by sunlight reflected from Earth.

Electromagnetic Radiation: The combined electric and magnetic fields that propagate as waves, including light and other forms of radiation such as radio waves, X-Rays, and Gamma Rays.

Electromagnetic Spectrum: The entire range of electromagnetic waves, from radio waves to gamma rays, sorted by wavelength or frequency.

Emitted Heat Radiation: The thermal radiation emitted by an object due to its temperature, often in the form of infrared radiation.

Fusion Reactor: A device that produces energy by combining atomic nuclei through the process of nuclear fusion, similar to the sun's energy generation.

Geomagnetic Storm: A disturbance in Earth's magnetosphere, often caused by solar activity, resulting in enhanced auroras and potential disruptions to technology and power grids.

Ionosphere: A region of Earth's upper atmosphere containing ionized gases, which affects the propagation of radio waves and is the location of the aurora phenomenon.

L1 Orbit: A point in space, known as a Lagrange point, where the gravitational forces of the Earth and the Sun balance, making it an ideal location for certain types of space observatories and missions.

Light Trespass: The unwanted or excessive illumination of an area by light, often causing light pollution.

Magnetosphere: The region around a planet or celestial body where its magnetic field interacts with the solar wind, protecting it from harmful radiation.

Overview Effect: A profound cognitive shift experienced by astronauts when viewing Earth from space, leading to a heightened awareness of the interconnectedness of life on our planet.

Phosphoresce: The emission of light or other radiation after a substance has absorbed energy and then re-emits it slowly over time.

Plasma: A state of matter in which electrons have been stripped from atoms, resulting in a highly ionized and electrically conductive gas.

Reflected Solar Radiation: Sunlight that is redirected off a surface or object, commonly observed in the Earth's energy balance.

Reflection: The process by which light or other waves bounce off a surface and change direction.

Refraction: The bending of light or waves as they pass from one medium into another, due to changes in speed.

Scattering: The redirection of light or other waves in various directions when it encounters particles or obstacles.

Wavelength: The distance between successive peaks (or troughs) of a wave, defining the wave's length.

Image Credits:

Cover: Proctor, Sian; SpaceX

1: NASA JSC

3: NASA JSC

4: NASA: background; Proctor, Sian: inset

5: White, Frank: inset

6-7: NASA

8: Proctor, Sian

10-11:NASA JSC

12-13: NASA JSC

14: Proctor, Sian

16-17: Proctor, Sian

18: NASA JSC

19: NASA

20: NASA

21: Adobe Stock: top; NASA: bottom

22:Adobe Stock

23: Adobe Stock

24: Adobe Stock

25: NASA MFSC: top left; Read, John A.: bottom right; da Vinci, Leonardo: bottom left

26-27: NASA

28-29: NASA

30: NASA JSC

31: NASA: left and right

33: NASA NOAA

34: NASA

35: NASA

36: NASA: background; NASA JSC: eclipse phases

37: Stellarium: background; NASA: inset

38-39: NASA

40: NASA

42-43: Read, John A.

44-45: Read, John A.: bottom background; Stellarium: background; NASA: moons

46: NASA

47: NASA

48: NASA

49: Adobe Stock

50-51: NASA: all

52:Adobe Stock

53: NASA

54: Bavarian State Library (public domain)

55: Musée Lugdunum (public domain)

56: Adobe Stock

57: Aublet, Albert (public domain): background; Unknown artist, The Metropolitan Museum of Art (public domain): inset

58: Adobe Stock

59: Nishimura Shigenaga (public domain)

61: Adobe Stock

62: Adobe Stock

63: Adobe Stock

64: Adobe Stock: background; NASA: insets

65: Proctor, Sian: background

66: NASA

68: NASA JPL

69: NASA

70: NASA: background, inset

71: NASA

72: NASA: left; Environmental Action Coalition Records: right

73: New York Times

74: NASA: far left and middle background, aqua, terra;

75: JAXA: left background, GOSAT; SNSA: odin; CSA: SCISAT; NASA: middle and right background

76: NASA

77: NASA

79: NASA

80-81: NASA GSFC

82-83: NASA

83: Adobe Stock: overlay

84-85: NASA

86: NASA

87: NASA GSFC

88: NASA JPL

89: NASA

90: NASA

91: NASA: background; CIA Factbook: inset

92: NASA

93: NASA

94-95: NASA

96-97: NASA

98: Proctor, Slan

100-101: NASA: background; Adobe Stock: astronaut

102-103: NASA

104: Adobe Stock: insets; NASA: background

105: NASA: background; NASA/RKK Energiya: left; Remo Timmermans: right

106: NASA JPL: background; NASA JSC: inset

107: NASA

108: NASA: background; NASA JSC: inset

109: NASA KSC: background; NASA: inset

110: NASA GSFC

111: NASA

112: NASA

113: NASA: background, inset; Space for Art: logo

114: Sharman, Helen: inset; NASA backgound

115: NASA

116: NASA

117: NASA

118: NASA

119: NASA

120: NASA

121: NASA

122: NASA KSC

123: NASA

124: Blue Origin

125: Blue Origin

126: NASA: background; Blue Origin: inset

127: Sembroski, Chris

128-129: SpaceX

130-131: NASA

134-135: NASA

136: NASA

137: NASA

138: NASA

139: NASA JPL

140: NASA

141: Read, John A.

142: Read, John A.

143: Read, John A.

144: NASA

145: Read, John A.

146-147: NASA

148: NASA

149: NASA

150: NASA

151: NASA JSC

152-153: NASA

154: NASA

155: NAA

156-157: NASA

158-159: Adobe Stock

160-161: Inspiration4

162-163: Inspiration4

164-165: Michael Whelan

166-167: NASA

168: NASA

169:NASA

www.ingramcontent.com/pod-product-compliance
Lightning Source LLC
Chambersburg PA
CBRC090247230326
41458CB00108B/6517